THE DIFFUSION OF THE REFORMATION IN SOUTH-WESTERN GERMANY, 1518–1534

by

Manfred Hannemann
Illinois State University

THE UNIVERSITY OF CHICAGO
DEPARTMENT OF GEOGRAPHY
RESEARCH PAPER NO. 167

1975

Copyright 1975 by Manfred Hannemann
Published 1975 by The Department of Geography
The University of Chicago, Chicago, Illinois

Library of Congress Cataloging in Publication Data

Hannemann, Manfred, 1938–
  The diffusion of the Reformation in southwestern Germany, 1518–1534.

  (Research paper–University of Chicago, Department of Geography; no. 167)
  Bibliography: p. 219
  1. Reformation–Germany, Southern. I. Title. II. Series: Chicago. University. Dept. of Geography. Research paper; no. 167.
  H31.C514 no. 167 [BR305.2] 910s [270.6'0943'4]
  75-14120
  ISBN 0-89065-074-8

Research Papers are available from:
The University of Chicago
Department of Geography
5828 S. University Avenue
Chicago, Illinois 60637
Price: $6.00 list; $5.00 series subscription

ACKNOWLEDGEMENTS

A study of the diffusion of the Reformation in Germany was first suggested to me by Marvin Mikesell, and received considerable impetus from student colleagues at the University of Chicago, of whom Yehoshua Cohen must be mentioned in particular. It soon became evident that sixteenth-century Germany was too large an area and too great a time-span. In order to make the topic manageable, it was necessary to limit research to a smaller region and time-period where and during which Luther's ideas enjoyed little official support. This confined me to the period from 1518 to 1534, respectively the beginning and end of the Reformation movement in southwestern Germany.

The field research for this study was conducted during the year 1971-72, and was supported by grants from the Deutscher Akademischer Austauschdienst. I wish to express my sincere gratitude for this interest and generosity.

While in Germany, I was fortunate to be affiliated with the Institut für Landesgeschichte at the university in Tübingen. I owe much to Professor Hansmartin Decker-Hauff, Elmar Kuhn, and Franz Quarthal of this institution for their interest and advice.

My appreciation goes also to Professors Chauncy Harris and Karl Butzer at the University of Chicago, who read the manuscript and offered suggestions that improved its content and style. I am also indebted to Christopher Müller-Wille for advice on the study's illustrations.

To Marvin Mikesell, my supervisor, I wish to express sincere thanks not only for suggesting this study, but also for encouragement and constructive criticism. I would also like to thank my sister, Siglinde Mitchell, for improving and typing a full draft of the manuscript. Finally, I am grateful to my wife, Maria, whose involvement in this effort has entailed constant support and numerous sacrifices.

TABLE OF CONTENTS

	Page
ACKNOWLEDGEMENTS	iii
LIST OF TABLES	viii
LIST OF ILLUSTRATIONS	ix

Chapter
 I. INTRODUCTION 1

        The Problem
        The Filtering System
        Personal Contact
        Spreading Luther's Ideas through Print and Writing
        Institutional Channels
        Successive Stages of the Reformation Movement
        The Role of Humanists and Clergy in the Early Diffusion of
            the Gospel
        The Sermon Movement
        The New Communion
        Abolition of the Mass
        Regional-Cultural Characteristics of Southwestern Germany
        The Relationship between Size of Town and the Occurrence of
            Preachers

 II. THE REFORMATION AND CONCOMITANT SOCIAL AND
    RELIGIOUS MOVEMENTS 41

        The Peasants' War
        The Relevance of Marxist Theory

 III. THE IMPACT OF HIGHER EDUCATION AND THE
     COMMUNICATION NET ON THE REFORMATION 55

        Universities and the Reformation
        Roads and other Communications
        Mining, Population Mobility, and the Reformation

 IV. POLITICAL DECISION-MAKING AND THE REFORMATION 67

        Imperial Cities
        Knightly Territories
        Princely Territories
        Ecclesiastical Territories

V. PLACES OF ORIGIN OF PREACHERS AND LOCALITIES WITH
RESIDENT PREACHERS 84

    Informal Diffusion
    Localities with Resident Preachers
        Territories of Ulm and Biberach
        The Danube Towns
        Constance
        Lindau
        Isny
        Cities on the Upper Rhine
        Upper Neckar Cities
        Middle Neckar Area
        Southern Franconia

VI. LOCALITIES WITH IMMIGRANT PREACHERS 119

    Itinerant Preachers
    Ulm Territory
    Interior Württemberg
    The Ries
    Eastern Franconia
    Main Basin
    North Baden
    Strasbourg Area
    The Breisgau

VII. PREACHING FOUNDATIONS (PRÄDIKATUREN) AND THE
DIFFUSION OF THE REFORMATION 146

    South Franconian Cities
        Schwäbisch-Hall
        Dinkelsbühl
        Nördlingen
        Ellwangen
        Heilbronn
        Weinsberg
        Mosbach
        Brackenheim
        Gemmingen
        Pforzheim
        Weil der Stadt
    Upper and Middle Neckar Region
        Stuttgart
        Waiblingen
        Reutlingen
        Göppingen
        Ehingen
        Rottweil
    Upper Rhine Area
        Offenburg
        Baden-Baden
        Lahr
        Heidelberg

       Main and East Franconian Cities
           Rothenburg/Tauber
           Würzburg
           Ansbach
           Windsheim
           Kitzingen
           Wertheim
       Ulm and Upper Swabia
           Ulm
           Giengen
           Dillingen
           Munderkingen
           Biberach
           Memmingen
           Isny
           Kaufbeuren
           Kempten
       Summary

VIII. CONCLUSIONS 209

BIBLIOGRAPHY 219

LIST OF TABLES

Table		Page
1.	List of Property Tax by Prefectures	52
2.	List of Property Tax by Administrative Town (Amtsstadt)	53
3.	Students from Various Cities Attending the Universities Heidelberg, Wittenberg and Freiburg, 1515-1524	60

LIST OF ILLUSTRATIONS

Figure		Page
1.	The Location of Southwest Germany within the Realm of the Holy Roman Empire	2
2.	Spatial Diffusion of Protestant Preachers and/or Ideas 1518-1524 .	10
3.	Spatial Diffusion of Protestant Preachers and/or Ideas 1525-1534 .	11
4.	The Age of Adopters at the Time of their First Sermon	19
5.	The Appearance of Lutheran Preachers 1518-1520	23
6.	The Appearance of Lutheran Preachers 1521-1523	24
7.	The Appearance of Lutheran Preachers 1524-1525	25
8.	The Appearance of Lutheran Preachers 1526-1530	26
9.	The Appearance of Lutheran Preachers 1531-1534	27
10.	Distribution of Communion under both Forms 1518-1534	30
11.	Localities which Abandoned Mass, 1518-1534	31
12.	Cities and Towns of Southwest Germany at about 1550	36
13.	Relationship between the Size and Number of Cities and Towns and the Year of Adopting a Preacher of the New Faith	39
14.	Major Roads and Thoroughfares of Southwest Germany	63
15.	The Political Division of Southwest Germany at about 1500	69
16.	Distribution of Preaching Foundations at about 1520	147
17.	Adoption Rate of Preachers 1518-1530	216
18.	Cumulative Adoption Curve of Luther's Ideas 1518-1530	216

CHAPTER I

INTRODUCTION

Among the most puzzling phenomena in cultural history are the occasional outbreaks and diffusion of "symbolic epidemics," including the origin and spread of a new religion or a new political ideology. Rarely has the spread of ideas characteristic of such movements been considered at the level of a total society.[1]

The Problem

An attempt is made here to describe and explain the diffusion of the Lutheran movement in southwestern Germany (Fig. 1) in the first fifteen years after its inception (1518 to 1534), a period which has often been regarded as the decisive era in the formation of the new church in Württemberg. By "movement" should be understood a gathering of people who were "moved" by and mutually dedicated to the propagation and implementation of the ideas, thoughts, and teachings of Martin Luther. This ideological movement resulted ultimately in a social movement, in which persons not only sought the company of existing adherents, but attempted to win new followers to their cause. The unfolding activities in this movement cast a vast influence on the political, economic, cultural, and social sectors of human existence for centuries to come.[2]

Since the movement had its roots in Martin Luther's 95 Theses of October 30, 1517, the adjective <u>evangelical</u> is synonymous with "Lutheran" in a broad sense.[3] It includes the thoughts and writings of theologians like Melanchthon,

---

[1] Kenneth E. Boulding, "The Learning Process in the Dynamics of Total Societies," <u>The Study of Total Societies</u>, ed. by Samuel Z. Klausner (Garden City: Anchor, 1967), p. 10.

[2] Hans Rösler, <u>Geschichte und Strukturen der evangelischen Bewegung im Bistum Freising, 1520-1571</u>, Einzelarbeiten aus der Kirchengeschichte Bayerns, No. 42 (Nürnberg: Verein für Bayerische Kirchengeschichte, 1966), p. 3.

[3] The term <u>evangelical</u> was applied to the movement by Luther personally in 1521, to demonstrate the biblicity of his teaching. At this stage, opponents of the Reformation denied Luther's supporters membership in the <u>one and only</u>

Fig. 1.--The Location of Southwest Germany within the Realm of the Holy Roman Empire.

Bucer, and Brenz, as well as those of persons and groups of various persuasions whose claims went in the same general direction.[1] Since the latter, variously called "Anabaptists," Schwärmer, and Schwenckfelders, appeared as separate movements only during the year 1525, and since research on them is already extensive, they are excluded from this investigation.

Epidemiological theorists and historians (including Marc Bloch) have drawn analogies between the successful spread of cultural phenomena and biological contagion. Once it has passed a critical threshold, a contagion will be successful and reach a large population, whereas failure to reach this threshold means that it will retract, either affecting only a small number of subjects or dying out completely.[2] However, what makes certain doctrines contagious is something about which we seem to have little understanding. Moreover, one must be careful in applying simple mechanistic interpretations to any diffusion study, for historic reconstruction on these premises can at best be only two-dimensional. In this respect, neither the Vienna School of Ethnology nor the Hägerstrand School, which considers diffusion from a geographical viewpoint, have been entirely successful in furthering understanding of why a material or, particularly, an ideological innovation meets acceptance or rejection. Both schools have failed to deal adequately with the third dimension inherent in any diffusion process, specifically the psychological or symbolic aspect of diffusion. It is interesting to note that this dimension is and has been the method by which opinion leaders recruit prospective followers.

The sustained transformation of an existing popular institution such as the Roman-Catholic Church, often requires necessary conditions that are antecedent and may even exclude a direct link to the goals that the founder of the ideology had in mind. In fact, these antecedents may have developed over a long

---

church, and applied labels like "Lutherans" or "Martinians" to them. The Catholics, in turn, were called "Papista." See Friedrich Lepp, Schlagwörter des Reformationszeitalters, Quellen und Darstellungen aus der Geschichte des Reformationsjahrhunderts, No. 7 (Leipzig: Heinsius, 1908), p. 485.

[1]See Peter F. Barton, "Das Jahr 1525 und die Abschaffung der Messe in Strassburg," in Reformation und Humanismus, Robert Stupperich zum 65. Geburtstag, ed. by Martin Greschat (Witten: Roth, 1969), p. 144, footn. 13; and Wilhelm Störmer, "Obrigkeit und evangelische Bewegung. Ein Kapitel fränkischer Landes- und Kirchengeschichte," Würzburger Diözesan-Geschichtsblätter, XXIV (1972), 115.

[2]Marc Bloch, The Historian's Craft (New York: Random House, 1953), p. 34.

period of time and at various levels of the social structure.[1] Objective and subjective factors that hastened the acceptance of the Lutheran movement included: (1) the conflict between the territorial power of bishops and their urban authorities; city governments and residents naturally opposed being subjected to ecclesiastical establishments over which they had no control; (2) the patronage factor, i.e., attempts of city administrations to seize power of their parish church from a monastery; (3) the image factor, such as the peasants' belief, although inaccurate, that Switzerland granted considerably more freedom to its citizens; and (4) the emerging nationalism that appealed to "Germans," and thereby creating a popular opposition to the papal "Italians." With the circulation and popularization of Luther's ideas, the latent religious conflict turned into a manifest one. Luther's attack resulted in a solidification and focalization of these and other antecedents into a full-blown movement against the all-dominant church hierarchy.

In geographical research the term "innovation" implies a synthesis of the spatial effect of the diffusion process and the psychological implications common in the reception or imitation of an idea. The spatial effect may not exhibit as close and direct a relationship to the physical landscape of a territory as to its connecting links, such as traffic, economy, settlement-pattern, social structure, and political organization, which are in themselves results of innumerable spatial and historical factors.

Of special significance is the role of urban centers and core areas that developed from increasing socio-economic differentiation through time. Urban centrality is measured or rated according to the surplus of supply, that is, the type and intensity of influences, services, and goods that radiate from a city into the countryside. It likewise includes the centripetal forces that converge on a city from its hinterland and <u>umland</u>.

Central place functions are not exclusive to one realm only. Besides the economic realm that has been emphasized in most studies, one can mention political-administrative centrality, cultural centrality, and ecclesiastical centrality. That these functions coincide in many cases goes without saying. However, the importance of cultural central places has received little attention. In the case of the Lutheran movement, smaller cities possessing cultural centrality in the form of a university adopted the new ideology at an earlier time than large neighboring metropoles. The new ideology received passionate and

---

[1] Edward A. Tiryakian, "A Model of Societal Change and its Lead Indicator," The Study of Total Societies, ed. Klausner, p. 75.

strong support in university centers, while localities without this support were dependent on surplus preachers. The influence of Mainz and Freiburg/Br. on Frankfurt and Strasbourg provides a good example (Fig. 2).

In addition to the influence of central place are singular functions that are based on such factors as the availability of minerals at a particular locality, or a favorable strategic location. The singular and supra-regional functions embodied in activities such as mining, industry, long-distance trade, and recreation, also favor the establishment of ties with cities and towns that may be far away from the places where such activity occurs. Finally, one has to take into account the personal connections between peoples of an umland, hinterland, or places farther away and one or several persons in a specific city or town.[1]

Personal connections or influences were of extraordinary significance in the diffusion of the Lutheran movement and hence deserve special attention in this study. Although central place theory may be of assistance in explaining some aspects of spatial distribution, a deeper explanation demands a consideration of additional problems: Who influenced a preacher to adopt the new ideology? Did he accept it in the position he formerly occupied as a Roman-Catholic or was he hired from outside as a messenger of the new faith? Which persons and institutions in what cities assisted him in his drive?

The Filtering System

The diffusion of an ideology such as the Reformation may correspond to an existing communication network, but this may be lacking where the new ideas were challenged by traditional circles or institutions defending the Roman-Catholic tradition. One might compare this communication barrier to a filtering system that promoters of the Reformation had to reckon with. The more advanced the movement, the more likely was interference by political authorities destined to act either to the advantage or disadvantage of the new ideology.[2] This interference was to be the cause of the decline of the movement both within and outside a locality or territory. It included the Pope at the top as well as the regular believers at the bottom of the Catholic hierarchy. In-between were the bishops, educated and uneducated clergy (sacerdotes literati and simplice),

---

[1] Klaus Fehn, Die zentralörtlichen Funktionen früher Zentren in Altbayern (Wiesbaden: Steiner, 1970), p. 2.

[2] J. D. Marte, Die Auswärtige Politik der Reichstadt Lindau von 1530 bis 1532, Beilage zum Jahresbericht der Kgl. Realschule Ludwigshafen a. Rh., No. 19 (Ludwigshafen: Ertel, 1904-05), p. 87.

church patrons, as well as political rulers and bodies. These persons and institutions pooled their energy to prevent any change in the direction of Lutheranism.

The prohibition from the top affected mainly preachers. Since they were by tradition obliged to their office, the adoption of the new doctrine was a proof of how deeply Luther's faith had affected them. We can therefore observe their defection from one territory into another, such as from Württemberg into Baden in the early 1520's, and in the reverse direction in 1531 and 1534. The same applies to evangelical preachers in Bavaria and Tyrol in the early 1520's, as they sought refuge in Upper Swabia and the Lake Constance area. The era of religious refugees, however, was still a century away, although Kenzingen and Rottweil with 150 and 400 religious refugees respectively in 1524 and 1529 provided a signal for later population shifts.

To change a religious dogma was a relatively easy task compared to the removal of the many deep-rooted social and economic arrangements already existing, such as property, possessions, revenues, and power structure. To reform these seemed to be an endless struggle, since the formal law always favored the old church.[1] The gradual introduction of the Reformation by rulers of their territories was the result of a complicated legal situation that often allowed a confessional change only after a Catholic priest had died. This is the reason why many communities were willing to hire evangelical preachers at their own expense "without any harm to the priest,"[2] who often continued in office until he retired.

A complete analysis of the information system is difficult if not impossible, since it would have to take account of the total amount of exchanged information, official as well as private, accidental as well as intended. However the available information offers some evidence of the ways in which the Reformation movement spread.

In the sixteenth century, three channels were available for the spread of the new Protestant ideology:

(1) Personal contacts, of direct or indirect nature.

(2) Personal and impersonal information spread by nonverbal means.

---

[1] Ibid.

[2] Johann Adam, <u>Evangelische Kirchengeschichte der Elsässischen Territorien</u> (Strassburg: Heitz, 1928), p. 201.

(3) Personal and impersonal information conveyed through institutional channels.[1]

To separate these three channels or networks would do violence to reality, since all three were interrelated. However, a few remarks on the effectiveness of each can be offered. It is important to remember that impersonal, or mass-communication, ushered in by the invention of movable type was and still is far less effective in symbolic conflicts than the communication process involving two or more persons in face-to-face contact.[2] Even personal letters fall short of this effect, though they are more effective than the printed page. Moreover, when we finally take into account the fact that the great majority of the population could neither read nor write, the importance of personal contact is manifest.

Personal Contact

In the category of direct personal contact, the following possibilities for relaying information exist:[3] (1) the source of influence and acceptor are sedentary and meet in their everyday life, or (2) the sedentary acceptor, as a member of a larger group, has the opportunity to learn to know the ideas of the originator. Both forms of idea transfer are based on word-of-mouth and can be summarized as "contact migration" or "diffusion by radiation."[4] In this process, the ideology diffuses in concentric innovation waves from the originator, provided that conditions favoring or delaying the idea are the same in all directions. Because of the condition of low personal mobility during the sixteenth century, this process was particularly adaptable to the lowest echelon of the

---

[1] Fritz Redlich, "Ideas. Their Migration in Space and Transmittal over Time," Kyklos, VI (1953-54), 301.

[2] Throughout the sixteenth century, the direct oral communication of business was preferred over letters and other types of written messages. Veit Bild of Augsburg, on the return of two of Luther's sermons to Bernhard Adelmann, preferred to give his opinion about Luther in personal meetings rather than in written form. Alfred Schröder, "Der Humanist Veit Bild, Mönch bei St. Ulrich," Zeitschrift des Historischen Vereins Schwaben und Neuburg, XX (1893), 203.

[3] Redlich, "Ideas," p. 302.

[4] Warren C. Scoville, "Minority Migrations and the Diffusion of Technology," Journal of Economic History, XI (1951), 349.

spatial hierarchical system and, therefore, was of only local significance. Although numerically this process affects the largest proportion of people, it corresponds with the urban-rural communication pattern and seldom overlaps the boundaries of a city or town and its <u>umland</u>. Had Luther's ideology been solely dependent on this type of idea transfer, it would have required decades to become the common property of Germans in the East, West, South, and North. And in all likelihood, the original content of the ideology would have been so disfigured after reaching more distant regions, that many of the ideas would have been difficult to recognize as Luther's.

A more realistic type of diffusion takes place with the actual migration of a man or of groups. It applied particularly to inter-urban communications that provided an extensive "rapid transit" network for the dissemination of Luther's ideas throughout Germany and beyond. Here, again, we have to differentiate among: (1) migration and diffusion of ideas by the originator himself; (2) a potential acceptor migrating to the source, absorbing the ideas and spreading them after returning to his homebase or elsewhere; and (3) diffusion of ideas by a migrating middleman, who connects the originator with another acceptor, be he again a middleman or the ultimate acceptor.

Since the originator of the ideology is usually its strongest advocate, his own missionary efforts and success outstrips that of any of his immediate followers or "secondary reformers." Luther's appearance in Heidelberg and Augsburg in 1518 and Worms in 1521 was of great significance to the diffusion of his ideas in southwestern Germany. At the Augustinian monastery in Heidelberg, where he appeared on April 26, 1518, he made an indelible impression on the younger humanists and clergymen, many of whom studied or taught at the local university. Luther's appearance before Cajetan in Augsburg in October of the same year was less successful, in spite of the courageous stand he took toward charges made by the papal legate. Since the "industrial capital" of Germany at that time lacked a university, the number of potential followers was small and relatively stationary. Both appearances, moreover, fell into a time-period when Luther had hardly any public appeal. Not until he had written <u>The Babylonian Captivity of the Church</u> and the <u>Open Letter to the German Nobility</u> in 1520, did he reach a large portion of persons in leading social, economic, and political positions. Luther's appearance before the Diet at Worms in 1521 had therefore special appeal to the political representatives from the various German principalities and cities of the Southwest, among them the Knight of Gemmingen, Count Georg of Wertheim, and Georg Vogler, the representative and secretary of the Margrave of Ansbach-Bayreuth. Upon their return home, they helped to further the cause of the Reformation by requesting Lutheran preachers from Wittenberg or other cities (Fig. 2).

The diffusion fostered by someone who migrates to the source and, equipped with the new ideology, returns to his hometown, falls into another category of indirect personal contact. It applied particularly to persons who came to the University of Wittenberg after 1518. When seen against the German scene as a whole, Wittenberg should be regarded as <u>primary</u> diffusion center, and Heidelberg, Augsburg, and Worms as <u>secondary</u> diffusion centers. However, from the perspective of southwestern Germany, the secondary diffusion centers might well appear as primary ones. Constance belongs to the same category, namely a secondary center within the German realm, and an important primary center in the southwestern German province. Strasbourg, Freiburg, Schwäbisch-Hall, and Esslingen belong in the tertiary rank of the diffusion scale (Fig. 3).

A modified version of this diffusion pattern comes into effect when the same acceptor secures a position for another preacher or opinion leader at his hometown or another city. Thus, Thomas Blarer studied in Wittenberg and recommended Lonicer as professor of Greek to Philipp Engelbrecht at the university in Freiburg/Breisgau.

The diffusion of the ideology by itinerant middlemen joining the originator, secondary, and tertiary reformers can be traced particularly to travelling salesmen, journeymen, and itinerant preachers. In spite of the scarcity of information concerning their activities, the following conclusions can be drawn: (1) businessmen and journeyman in particular had a share in making the populace aware of the existence of the new ideology and (2) this awareness led many communities to demand a preacher of the new faith.

Spreading Luther's Ideas through Print and Writing

The second main channel for spreading Luther's ideas lay in the distribution of books, manuscripts, pictures, personal letters, and songs. The distribution center for most of the second-hand information was Frankfurt/Main, where the sale of Reformation pamphlets and other printed material yielded high profits at the semi-annual fair.[1] Here, booksellers from Leipzig handed

---

[1] Heinrich Steitz, <u>Geschichte der evangelischen Kirche in Hessen und Nassau, I: Bewegung, Reformationen und Nachreformationen</u> (Marburg: Trautvetter & Fischer, 1962), pp. 6, 22.

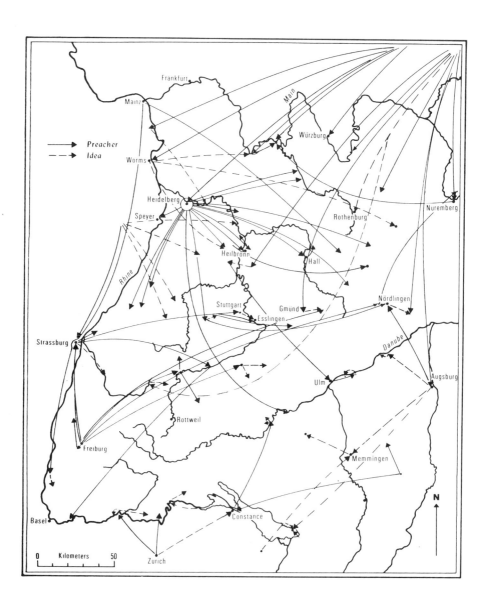

Fig. 2.--Spatial Diffusion of Protestant Preachers and/or Ideas 1518-1524

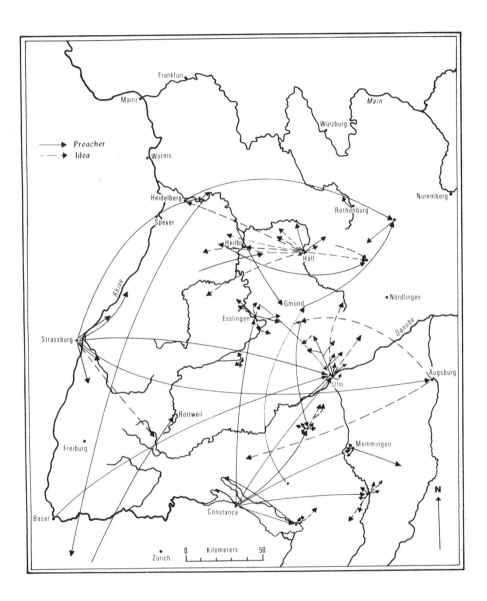

Fig. 3.--Spatial Diffusion of Protestant Preachers and/or Ideas 1525-1534

out Lutheran literature to attending booksellers and printers from Augsburg, Nuremberg, Basle, and other cities, who, in turn, reprinted it with the moral support of local intellectual circles. From these publishing centers the literature was distributed to other cities and towns in southwestern Germany and distant areas. By these means, Lutheran ideas were disseminated often without any closer personal contact. Mere physical distance was not as important as the size of a city, so that a typical case of diffusion would be from the largest cities at the top down to the smaller towns and villages. In this respect, the hierarchical diffusion of printed material was often dissimilar to the diffusion encouraged by personal contact.

Most of the literature was authored and distributed by persons who belonged to the intellectual elite, such as Martin Bucer, Urbanus Rhegius, Christoph Schappeler, Joachim Vadian, Nikolaus Gerbel, and others.[1] Some of them hired a <u>colporteur</u>, whose sole task was to sell and distribute Luther's works as well as satirical literature directed against the Catholic Church.[2] The cities and surrounding areas that hosted most of the Lutheran preachers were also diffusion centers for this literature.[3] Thus, the Basle printer Adam Petri, who notified Zwingli about the newly-published brochures of Luther, asked the reformer to recommend them not only to his parish members but to other preachers as well.[4]

The high rate of illiteracy was not limited to the peasantry but included a large segment of the urban population as well, thereby limiting the effectiveness of the written word on middle-class burghers and patricians. The use of visual devices such as woodcuts and copperplates, as well as oral communication partially compensated for this information-gap. With the help of these communication media, and in concert with various vocal supporters, the common man often began to show serious interest in Luther's teachings. Even as late as the sixteenth century, the Basle magistracy issued its decrees not in posters but

---

[1] Paul Kalkoff, "Die Stellung der deutschen Humanisten zur Reformation," Zeitschrift für Kirchengeschichte, XLVI (1928), 176.

[2] Paul Kalkoff, "Jakob Wimpfeling und die Erhaltung der katholischen Kirche in Schlettstadt," Zeitschrift für die Geschichte des Oberrheins, N.F. XIII (1898), 113.

[3] Karl Müller, Kirchengeschichte, Vol. I, Pt. 2 (Tübingen: Mohr, 1904), p. 269.

[4] Jakob Mähly, "Beatus Rhenanus," Beiträge zur Vaterländischen Geschichte der Historischen Gesellschaft zu Basel, VI (1857), 187.

through town criers.[1]

Institutional Channels

The diffusion of evangelical notions in southwestern Germany also occurred through institutional means that already existed prior to the birth of Lutheranism. Such connections or associations either were the basis for, or supplemented and reinforced personal contacts. Where institution and person merge, we can expect a multiple effect that is difficult to sort out and separate.

Universities in particular were in a position to influence the course of events in far-away cities and towns. Professors, teachers, and students, during and after their studies, frequently returned to their home-base as preachers, officials, and private citizens to stand up for the gospel. The examples of Luther, Oecolompadius, Engelbrecht, Brenz, and others at the universities in Wittenberg, Basle, Freiburg, and Heidelberg, best illustrate the unique and important role of persons and institutions in the spread of the new dogma. In spite of his active preaching schedule in the small rural town of Wittenberg, Luther would have remained only a figure of local interest had it not been for his association with the university.[2] The fame of Wittenberg University, in turn, rested on its two leading theologians, Luther and Melanchthon.[3] Subjects other than theology could be studied as well or better in the older and well-established universities located in the west and south of Germany. Needless to say, Luther first shared his thoughts and perceptions with selected members of the Wittenberg theological faculty, before sharing them with his students and the general public.[4]

Also important for the diffusion of the evangelical creed was the contact between the administrations of the imperial and territorial cities and their uni-

---

[1] Hans Wackernagel, "Volkstum und Geschichte," Basler Zeitschrift für Geschichte und Altertumskunde, LXII (1962), 173; and Gottfried Seebass, "Die Reformation in Nürnberg," Mitteilungen des Vereins für die Geschichte der Stadt Nürnberg, LV (1967/68) 256.

[2] Kurt Aland, "Die Theologische Fakultät Wittenberg und ihre Stellung im Gesamtzusammenhang der Leucorea während des 16. Jahrhunderts," in Kirchengeschichtliche Entwürfe (Gütersloh: Mohn, 1960), p. 284; and Bernhard Klaus, "Herkunft und Bildung lutherischer Pfarrer der reformatorischen Frühzeit," Zeitschrift für Kirchengeschichte, LXXX (1969), 43.

[3] Aland, "Die Theologische Fakultät Wittenberg" (1960), p. 313.

[4] Ibid., p. 306.

versities. Cities such as Heilbronn, Schwäbisch-Hall, and Ansbach obtained their preachers and schoolmasters by contacting the university in Heidelberg (Fig. 2). In these instances, the migration of ideas becomes institutionalized to its fullest extent: Men migrate between institutions with new ideas in their heads. What happened between various institutions of various cities also occurred on a more local scale between cities and neighboring villages. Once convents and monasteries in a certain city were forced to switch to the new faith, their possessions, often in the form of village churches outside the city, likewise became Protestant.

Successive Stages of the Reformation Movement

The Reformation transformed an existing ecclesiastical orientation and organization into a new and different one. It clearly was a gradual procedure that depended upon the support of leading political circles. In this respect, the Reformation might be compared to Weber's "voluntary association" (Verein), a communal group originating through voluntary agreement that developed out of a "compulsory association" (Anstalt). In this analogy, the Roman-Catholic Church, with its authoritarian imposed laws, belongs to the latter category. During their initial period the emerging Protestant groups were held together primarily by a personal act of adherence.[1]

A religious associative relationship, be it sect or confession, can be created in one of two ways: (1) by voluntary or spontaneous action "from below," or; (2) by being imposed "from above."[2] These terms should not be interpreted in sociological class terminology but from an action-typological (wirkungstypologisch) viewpoint.[3]

Traditionally, the Reformation has been divided into (1) a religious reformation and (2) a political reformation. The religious reform movement originated and found support among the educated clergy and leading personalities in city and territorial governments. They vied for popular support from below, but also tried to obtain recognition and support from the political ruler

---

[1] Max Weber, The Theory of Social and Economic Organization, ed. by Talcott Parsons (Glencoe: Free Press, 1964), p. 151.

[2] Ibid., p. 148.

[3] Peter Schöller, "Kräfte und Konstanten historisch-geographischer Raumbildung," Festschrift für Franz Petri zum 65. Geburtstag (Bonn: Röhrscheid, 1970), p. 479.

or ruling bodies. The movement was initiated by Luther's 95 Theses on October 30, 1517, as well as by two published sermons about indulgence and grace that repeated the Theses in greater detail. Since the latter were published in German, they could be read and further distributed by literate laymen and members of learned circles. Often unable to understand the original Theses, which were in Latin, these groups gained a clearer picture of Luther's argument.[1] The rapid diffusion of the argument all over Germany and beyond must be ascribed largely to this document.[2]

In the early sixteenth century, the clergy and other members of the intelligentsia still held absolute monopoly over the dispersion of information. There is little doubt that the agitation of the common people for religious reform came chiefly through them.[3] They zealously followed the development of the Lutheran movement by keeping contact with friends who were closer to the original source, and then promoted the new doctrine among their less enthusiastic friends and neighbors who looked to them for the interpretation of distant and unfamiliar events.[4] Distance had a divisive effect primarily because of the imperfect means of communication. The area of personal familiarity was more confined then than it has been in modern times. Even so, widening trade contacts resulted in an expansion of communication. Growing consumer demands led to an increased market potential in cities and large towns, and a corresponding decline of the significance of smaller towns and isolated regions.[5]

Just as information passes more easily from higher status sources to lower ones, Luther's message filtered down from the top, steadily expanding its base in an avalanche-like manner. Accordingly, the bishop of Augsburg complained to the Memmingen magistrate in July 1523 that among its citizens "a small number of unlearned laymen are instructed in the teachings of Luther,

---

[1] J. Luther, Vorbereitung und Verbreitung von Martin Luther's Thesen, Greifswalder Studien zur Lutherforschung und neuzeitlicher Geistesgeschichte, No. 8 (1933), p. 24.

[2] Schröder, "Der Humanist Veit Bild," p. 202; and Joseph Lortz, The Reformation in Germany (2 vols.; New York: Herder and Herder, 1968), I, 232.

[3] Hajo Holborn, "The Social Basis of the German Reformation," in The Reformation: Material or Spiritual? ed. by Lewis W. Spitz (Lexington: Heath, 1962), p. 41.

[4] Karl W. Deutsch, The Analysis of International Relations (Englewood Cliffs: Prentice-Hall, 1968), p. 103.

[5] Heinrich Kramm, "Landschaft und Raum als ökonomische Hilfsbegriffe," Vierteljahrschrift für Sozial- und Wirtschaftsgeschichte, XXXIV (1941), 11-12.

and are said to have instructed a great number of others in turn."[1] That Luther's language and originality were appealing to the majority of Germans can not be denied, although it should be realized that the masses were guided by more intelligent opinion leaders who could read and write and were able to reinterpret Luther's ideas within a local context. To interpret the evangelical movement as having sprung primarily from the poorer urban classes is simply not acceptable, as will be shown later on.[2]

The Role of Humanists and Clergy in the Early Diffusion of the Gospel

As Moeller's interesting study points out, the importance of the already existing humanist circles for the diffusion of the 95 Theses cannot be underestimated.[3] They found their way from one humanist circle to another, and it is largely to the credit of these groups that the Reformation, even against Luther's intentions, found its way to other parts of Germany, mainly the Southwest, where humanism was firmly entrenched. In Augsburg, the first Lutheran preachers emerged and moved from the rather confining humanist environment to the market place. Influential popular preachers without a humanist orientation emerged somewhat later.

The individualistic, subjective, and "democratic" ideas of the humanists implied an autonomization of the individual that led to sharp criticism of the Roman-Catholic Church. Rather than place emphasis on the objective worship cult, the humanists turned inward and attempted to live and act according to the Holy Scripture.[4] Their predilection for original sources as well as their condemnation of the traditionally conceived contrast between laymen and professionals in religious and secular affairs, closely conformed to Luther's reemphasis of Scripture.[5] Unlike a large section of the urban and rural population, they

---

[1] Friedrich Dobel, *Memmingen im Reformationszeitalter nach handschriftlichen und gleichzeitigen Quellen* (Memmingen: Besenfelder'schen, 1877), p. 32.

[2] This mistaken view has been expressed most recently by Störmer, "Obrigkeit und evangelische Bewegung," p. 123.

[3] Bernd Moeller, "Die deutschen Humanisten und die Anfänge der Reformation," *Zeitschrift für Kirchengeschichte*, LXX (1959), 46-61; esp. p. 50.

[4] Matthias Simon, "Die Stiftspredigerstelle zu Öhringen als Movendelpfründe," *Würzburger Diözesangeschichtsblätter*, XXVI (1964), 190.

[5] Georg Kaufman, *Geschichte der deutschen Universitäten* (2 vols.; Stuttgart: Cotta, 1896), II, 523. Moeller, "Die deutschen Humanisten und die Anfänge der Reformation," p. 53.

embraced the new teaching in pure form, without the added force generated by political and social conflicts. This attempt to "democratize" a movement has been underestimated or overlooked; yet it is an important clue to the success of the Reformation.[1]

Luther was not accepted by all humanists, and some famous ones, such as Erasmus, Fabri, Johann von Botzheim, Zasius, and Pirkheimer, soon withdrew their support. Others such as Bild followed them in 1524-1525, with the onslaught of the Peasants' War, which was blamed on Luther. These men emphasized the need for reforming the entire church and not merely the development of a separate church or sect.[2]

Ironically, their attitude had little effect on the already immensely popular movement. Luther's ideas inspired first of all those younger humanists who were well-educated in the classics. These literates and poets stood for free and independent research and interpretation of the Bible. As the papal legate in Germany, Aleander, indicated: "They [grammarians and poets] only consider themselves intellectuals by deviating from the accepted path of the church." He also referred to the Greek poets as "Luther's accomplices."[3]

The common folk, peasants, craftsmen, and even many of the better-to-do burghers and clergy, frequently lacked a genuine interest in and understanding of ideological change.[4] Therefore, they followed the call of religious and political opinion leaders who acquainted them with the new ideas. Without doubt, the Lutheran reputation with all its concomitants preceded the Lutheran doctrine which came to the public in the shape of sermons and liturgical changes. When peasants who referred to Luther's "evangelical freedom" were asked by opponents about its meaning, they could only allude to the Wittenberg reformer. It is significant that to the majority of Germans, Luther was better known as a man of action, a public figure, than a theologian and academician.[5] Until the

---

[1]Peter Albert, "Die reformatorische Bewegung zu Freiburg bis zum Jahre 1525," Freiburger Diözesan-Archiv, XIV (1919), 4.

[2]Schröder, "Der Humanist Veit Bild," p. 188.

[3]Peter Albert, "Die reformatorische Bewegung zu Freiburg," p. 4. Melanchthon, Unger, Phillip Engelbrecht, Lonicer, Vadian, Gerbel, von der Busche, Grynäus, Münster, Agricola all personified this movement.

[4]Ernst W. Zeeden, "Grundlagen und Wege der Konfessionsbildung in Deutschland im Zeitalter der Glaubenskämpfe," Historische Zeitschrift, CDXXXV (1958), 271.

[5]Hans J. Hillerbrand, "The Spread of the Protestant Reformation of the

late 1520's, the epithet "Lutheran" was synonymous with many of today's derogatory descriptions, such as "radical," "hippie," or "commie."[1] It is not surprising, then, that for the common people the Wittenberg reformer was a symbol of opposition to the established church and all it stood for.[2]

When Luther made his Ninety-five Theses public he was about 35 years old, still a young man. The vast majority of his adherents were still younger than the Wittenberg reformer (Fig. 4), whereas his Catholic opponents, with the exception of Eck were older. In other words, a definite generation gap existed between these two factions. The conversion of the common man was only cursory, with real participation and interest developing only in association with devoted, young opinion leaders and, later, institutionalization.[3] In Basle, the earliest reform attempts were marred by a period of disquiet and unrest that cried for a clear articulation of the religious issue and an indigenous religious reformer who could exploit the new situation.[4] This explains to some degree Luther's condonement by several Catholic bishops prior to the "get tough" directions they received from above.

While the opponents of the evangelical movement blamed its followers for the outbreak of the peasants' rebellion, the adherents of Luther turned the argument around and blamed the continuation of the Latin mass and the monasteries.[5]

Sixteenth Century: A Historical Case Study in the Transfer of Ideas," South Atlantic Quarterly, LXVII (1968), 283.

[1] According to Dr. Robert Fischer, Lutheran School of Theology, Chicago. Even decades later, the activities of Lübeck Lutherans were described condescendingly as "eating meat, contempt for churches, images and sacraments, creating agitation." Gerhard Pfeiffer, "Das Verhältnis von politischer und kirchlicher Gemeinde in den deutschen Reichsstädten," in Staat und Kirche im Wandel der Jahrhunderte, ed. by Walther P. Fuchs (Stuttgart: Fischer, 1966, pp. 79-99), p. 88.

[2] The strong influence of Luther stems less from the content of his 95 Theses than the news of his opposition or protest. The disenchantment of the multitude with the system was so marked, that in the long run his impact as a person outweighed his religious and ideological thrust. See Gottfried Müller, "Reformation und Information," in Die Bedeutung der Reformation für die Welt von Morgen, ed. by Rainer Schmidt (Frankfurt: Lembeck, 1967), p. 204; and Prof. Brian Gerrish, University of Chicago, lecture notes.

[3] Otto Kähni, "Reformation und Gegenreformation in der Reichsstadt Offenburg und Landvogtei Ortenau," Die Ortenau, III (1950), 27.

[4] Rudolf Wackernagel, Geschichte der Stadt Basel, III (Basel: Helbing und Lichtenstein, 1968), 330.

[5] Rudolf Herrman, "Die Prediger im ausgehenden Mittelalter und ihre Be-

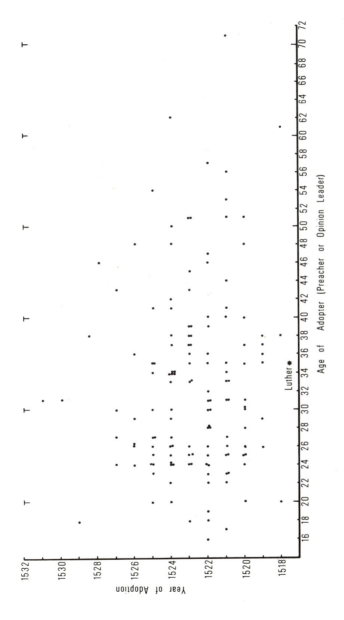

Fig. 4.--The Age of Adopter at the Time of their First Sermon

However, this counter-charge regarding the mass is not completely acceptable. While the peasants hardly attacked the Latin mass, they did attack cloisters and members of monastic orders. They protected and liberated secular priests who supported Luther's teaching, such as Johann Mantel of Stuttgart and Jost Höflich of Ulm.[1] By the same token, Johannes Zwick was hailed as herald of freedom throughout Upper Swabia.

True to its ecclesiastical origin, the Reformation took place in three characteristic stages in the following sequence: It began with (1) the evangelical sermon or gospel reform, which in many places turned into (2) reformation of the liturgy, and, finally, ended in (3) constitutional reform that lent a firm organizational framework and legitimized the entire movement preceding it.[2] The first stage was accomplished in a fairly short time, while each succeeding stage demanded a longer period of time. Thus, it took the sermon movement an average of five years to spread over the country, while the change of communion under both kinds, and the abolition of the mass took approximately seven and nine years, respectively.

During the first decade of the Reformation, the true inclination of the populace became apparent by their acceptance of the gospel and, specifically, their rejection of communion sub unie specie (under one kind), and the mass.[3] In almost all southwest German localities where the Reformation was successful, these changes were carried out in succession. Not until the year 1525 were the latter two considered to be of equal importance. Moreover, only the courageous members of the clergy ventured to gradually abandon these traditional

---

deutung für die Einführung der Reformation im Ernestinischen Thüringen," Beiträge zur Thüringischen Kirchengeschichte, I (1929/31), 21.

[1] Johannes E. Pfister, Denkwürdigkeiten der Württembergischen und Schwäbischen Reformationsgeschichte (Heilbronn: Renner, 1938), pp. 22-23.

[2] Martin Heckel, "Reformation: Rechtsgeschichtlich," in Evangelisches Staatslexikon, ed. by Hermann Kunst et al. (Stuttgart: Kreuz, 1966), col. 1812. The increasing rigidity of the intellectual atmosphere created by the codification was opposed by a number of Protestant leaders on the same ground they rejected the institutionalized atmosphere of the traditional Roman-Catholic establishment. See: Hermann Buck, Die Anfänge der Konstanzer Reformationsprozesse, Österreich, Eidgenossenschaft und Schmalkaldischer Bund 1510/22-1531, Schriften zur Kirchen- und Rechtsgeschichte, No. 29-31 (Tübingen: Osiander, 1964), p. 134.

[3] Gerhard Katterman, Die Kirchenpolitik Markgraf Phillip I von Baden (1515-1533), Veröffentlichungen des Vereins für Kirchengeschichte in der evangelischen Landeskirche Badens, No. 11 (Karlsruhe: Besau, 1936), p. 35.

ceremonies.[1] It should be kept in mind that the abolition of the mass meant a total break with the traditional Roman-Catholic establishment. It is therefore considered a key indicator for the official introduction of the Reformation, that is, the political reformation. Whereas until then the Roman-Catholic and Protestant church services could be and often were performed in one and the same church, the former was now to be banned altogether.

The Sermon Movement

Luther's ideas were disseminated and accepted in cities and villages in the form of sermons. As far as ecclesiastical reorganization was concerned, the Wittenberg reformer relied on the power of the gospel to move the communities and parishes.[2] He stressed that "neither seal, decree, custom nor any law" must prevent the proclamation of the gospel. Since the burghers built and financed their own parish church, Luther argued that they should have the right to select appropriate preachers.[3] His constant emphasis on the pure proclamation of the Word points to the pulpit as prime means for carrying out the Reformation.[4] In addition, the acquisition of an educated preacher who could competently explain and interpret the Bible was one of the most desirable achievements in the early movement. Until the 1520's, the appointment of a clergyman rested strictly with the ecclesiastical authorities associated with the bishop. Their refusal to appoint an evangelical minister caused the parishes to select one themselves and ask the magistracy to confirm him. This innovation was accomplished with the aid of leading reformers in many cities of southwestern Germany. Traditional connections between towns, as well as newly-established personal connections based on the new faith proved to be useful in strengthening a movement that was still in the beginning stage. Localities with a more advanced movement assisted neophytes with theological as well as practical ad-

---

[1] Traugott Schiess, ed., Briefwechsel der Brüder Ambrosius und Thomas Blaurer, 3 vols. (Freiburg: Fehsenfeld, 1908-12), I, 165.

[2] Alfred Schultze, Stadtgemeinde und Reformation: Recht und Staat in Geschichte und Gegenwart, No. 11 (Tübingen: Mohr, 1918), p. 39.

[3] Ulrich Stutz, "Luthers Stellung zur Inkorporation und zum Patronat," Zeitschrift der Savigny-Stiftung für Rechtsgeschichte, Kan. Abt., XXXII (1911), 312.

[4] The preacher had a strong influence on the thinking of his parish members. For this reason, in the beginning of the Reformation movement, the entire emphasis was placed on the proclamation of the Word of God. Accordingly, the title of the preacher was Verbi Divini Minister, or "Servants of the Bible."

vice and the selection of preachers.[1] The leading reformers were especially eager to render their services to other cities, as will be seen later in the case of Blarer, Oecolompadius, Schappeler, Bucer, Eberlin von Günzburg, Wanner, Brenz, and others. It appears that those preachers who were most influenced by others were frequently younger and less experienced. Converts in villages and towns that had no evangelical preacher of their own attended sermons and enjoyed communion under both forms in neighboring towns.

Most, if not all maps that show the distribution of the Reformation movement in cities and other territories, are concerned with the "complete" Reformation, i.e., the political reformation. They are based on data from the 1540's or later, long after Luther's theology stirred up a sizable segment of the population. This approach overlooks cities whose citizens had already rallied to the new faith, but whose local and regional governments arrested the movement. To gain a more correct impression of the Reformation movement, it was necessary to look at the distribution and date of the first Protestant ministers (Figs. 5, 6, 7, 8, 9).

There were various kinds of preachers: clandestine, itinerant, regular, and foundation preachers (Prädikanten). In some localities they appear in this sequence, although no city or town displayed the entire hierarchy. Princes who had embraced the new faith usually kept a court preacher. In any case, all these preachers proclaimed the Word of God after being requested to do so by their respective parishes.

No other representative of the new faith had as much contact with the broad majority of the public as the preacher. He was the continuous source of information and his sermons were such an excellent and frequently-used propaganda tool that the entire confessional quarrel appeared as a gigantic preaching dispute. Yet, not all preachers took the side of one party. As is common in transition periods, there were a few who preached the evangelical way at one time and the Catholic way at another. The public had a name for these: Beiderhänder or "ambidextrous preachers."

A few remarks about the different types of preachers may be useful. All considered themselves as qualified heralds of the gospel. Clandestine and itinerant preachers (Winkel- and Wanderprediger) were mostly non-residents in their sphere of influence and used a pseudonym so they could operate in secrecy and appear as equal in the socio-economic view of the common people. This made them more popular among the common folk. The foundation preacher

---

[1] Alfred Schultze, Stadtgemeinde und Reformation, p. 39.

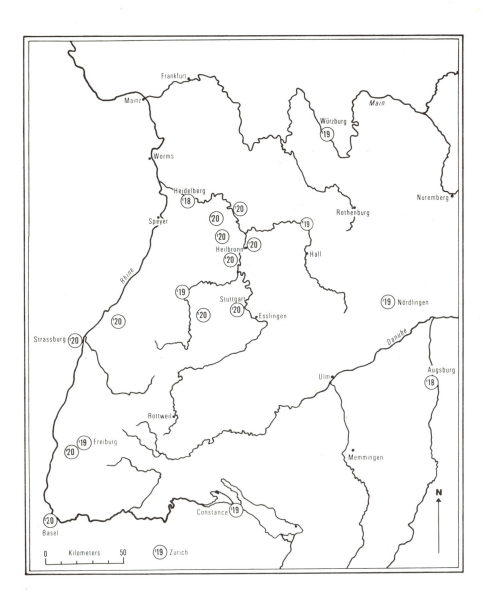

Fig. 5.--The Appearance of Lutheran Preachers 1518-1520

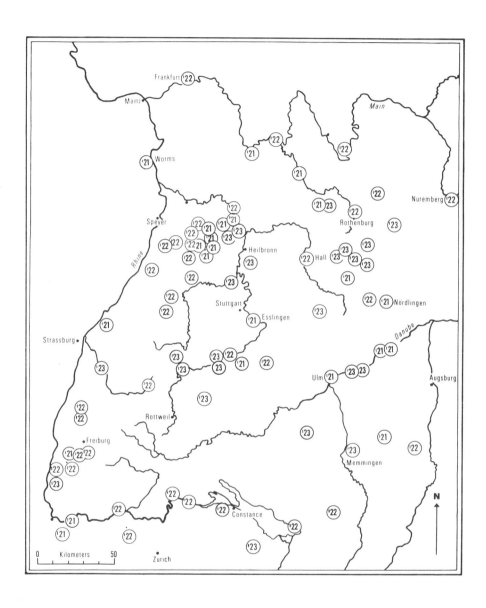

Fig. 6.--The Appearance of Lutheran Preachers 1521-1523

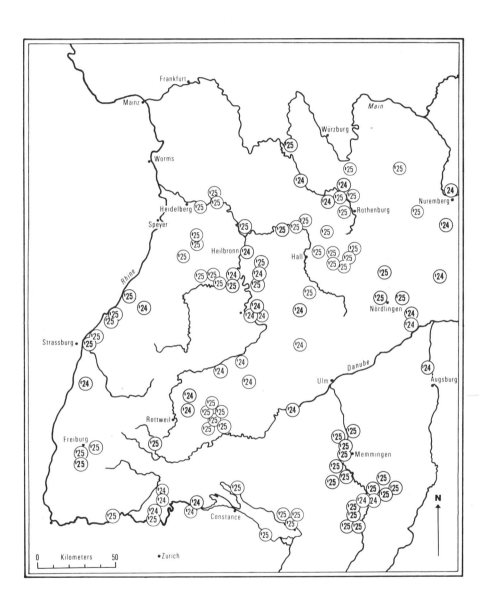

Fig. 7.--The Appearance of Lutheran Preachers 1524-1525

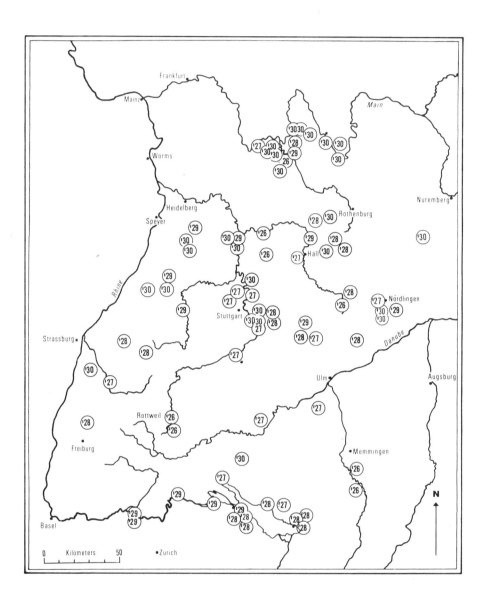

Fig. 8.--The Appearance of Lutheran Preachers 1526-1530

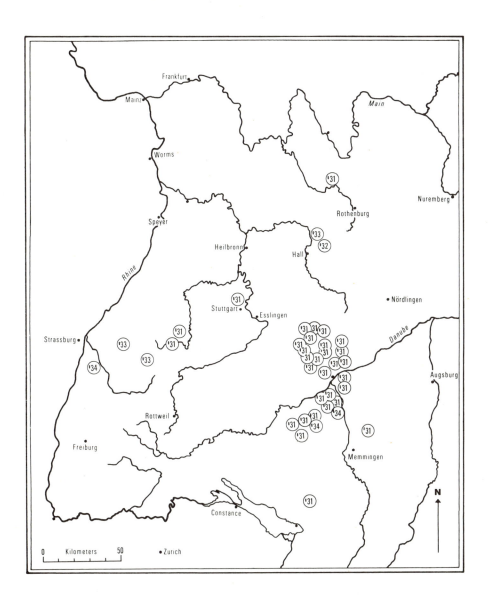

Fig. 9.--The Appearance of Lutheran Preachers 1531-1534

(Prädikant) differed from the regular preacher in that his task usually consisted of preaching the gospel and nothing else. Since almost all foundation preachers were required to have a good academic background, they surpassed the regular preacher intellectually and in prestige. Unlike the office of the regular preacher, the preaching foundation (Prädikatur) was a typical urban phenomenon.

The New Communion

As mentioned earlier, the issue of the communion was only of secondary importance in the beginning of the Reformation. In many places, for example in Dinkelsbühl, Baden-Baden, and Gemmingen, the communion was administered in the same church, sometimes by the same preacher, in both the sub une specie and sub utraque specie[1] forms, until the middle 1520's. This was a temporary solution in case of conflicting interests, the patron advocating the old line and the majority of the parish accepting the new faith.

In 1525, the Zwinglian-Lutheran dispute about the nature of the communion made its impact on the cities in southwestern Germany. The Lutherans maintained the real presence of Christ in the Lord's Supper and the Zwinglians insisted that the ceremony had symbolic value only. Whereas the Roman-Catholic Church believed in Transsubstantiation, the Lutherans emphasized Consubstantiation.

Abolition of the Mass

The mass was the most sacred of the traditional sacramental ceremonies. It was deeply ingrained in the popular piety and was therefore the last and most important barrier in breaking with the Roman-Catholic Church.[2] According to Nikolaus Manuel of Berne, it symbolized "the rock in the swamp"; for the serious-minded reformers it figured as a shield that protects "man-made doctrines and rules." Economically it was of great importance, since a large part of the ecclesiastical property came from mass foundations. The majority of the cler-

---

[1] Communion sub unie specie refers to the Roman-Catholic practice of serving only bread, whereas sub utraque specie refers to the Lutheran way of serving both bread and wine.

[2] Oecolompadius complained in 1522 on the Ebernburg that most people cling to the daily mass, in which "they hear the unintelligible murmur and see ceremonies without gaining anything." See Wolfgang Jung, "Zur Geschichte des evangelischen Gottesdienstes in der Pfalz," Veröffentlichungen des Vereins für pfälzische Kirchengeschichte, VII (1959), 7.

gy depended on the latter for their income, and this partially explains the great passion with which this battle was being fought. The mass was abolished only after the evangelical movement had received the endorsement of the absolute majority in the various city governments, which were, in turn, under pressure from the growing Protestant community.[1] Some clergymen approved of Luther's teachings, but had reservations about changing the sacramental system, while others placed such pronounced emphasis on the abolition of the mass that they left their town to find a position where this last vestige of Roman-Catholicism had been eliminated.

The day before the mass was officially outlawed in Zurich and Biberach, numerous believers took the opportunity to attend the mass and receive the sacrament in the traditional way, a sign that the magistracy had not abolished a sacrament that had become unpopular. As has been stated before, a change in belief was a comparatively easy task when measured against the change in and removal of long-existing liturgical rites such as the mass. The large number of localities with preachers (Figs. 5, 6, 7, 8, 9) in contrast to the relatively few places that instituted the new communion (Fig. 10) and abolished mass (Fig. 11) illustrates this fact. Consequently, the abolition of the latter not only demanded a longer period of time, but also depended on expert advice from lawyers in the large metropoles. As it happened, those cities that had already gotten rid of the mass pressured and advised smaller cities to follow suit.[2]

In the Zwinglian-oriented cities and towns of the South, and also in Esslingen, Ulm, and Augsburg, the abolition of the mass gave the signal for the removal or destruction of images in churches and chapels. In the Lutheran-oriented cities further north, the mass was not strictly abolished. Instead, the most offensive parts, such as the canon, were left out, so that Luther's more conservative followers or "anxious minds" would hardly recognize any change at all.[3]

---

[1] Ernst-Wilhelm Kohls, "Evangelische Bewegung und Kirchenordnung in oberdeutschen Reichsstädten," Zeitschrift der Savigny-Stiftung für Rechtsgeschichte, Kan. Abt., LXXXIV (1967), 131.

[2] Werner Näf, Vadian und seine Stadt St. Gallen, 2 vols. (St. Gallen: Fehr'sche Buchhandlung, 1957), II, 278.

[3] Erwin Iserloh, Der Kampf um die Messe, Katholisches Leben und Kämpfen im Zeitalter der Glaubensspaltung, No. 10 (Münster: Aschendorffsche, 1952), p. 7.

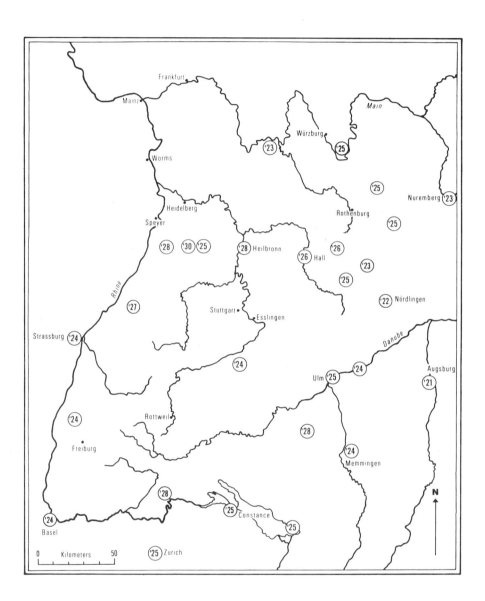

Fig. 10.--Distribution of Communion under both Forms 1518-1534

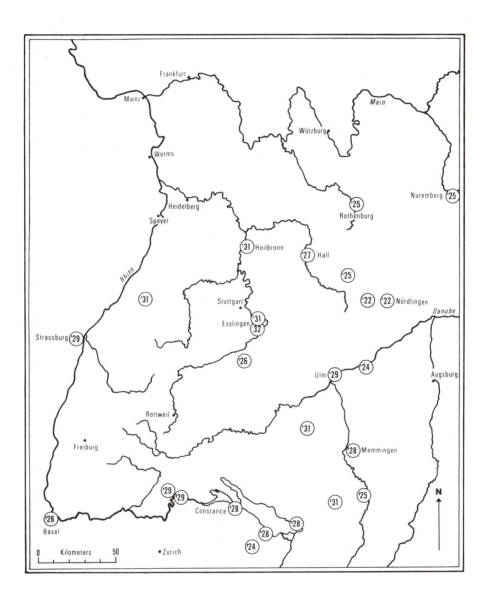

Fig. 11.--Localities which Abandoned Mass 1518-1534

Regional-Cultural Characteristics
of Southwestern Germany

Sixteenth century southwestern or Upper Germany was the most central and viable part of the German core area in an economic, political, and social sense. The latter was bounded by the city of Cologne in the northwest, Nuremberg and Augsburg in the southeast, and Basle in the southwest. This Upper German area was bounded by the upper Rhine in the west and south, the Main river in the north, and the Bavarian electorate in the east.[1] The comparatively large duchy of Württemberg comprised the central area of this region, an area characterized by a great number of juxtaposed and interpenetrating political entities (Fig. 15). This region could boast of a fairly dense population and a degree of urbanization matched by no other part of the Holy Roman Empire. Nowhere else could so many imperial city enclaves be found in such a limited area. Since this was more or less the center of trade and industry, and the image or aura of the empire was nowhere as vivid as in this section of Germany, it simply regarded itself as "the Empire."[2] Hence, it functioned as a crutch for many of the mini-states whose territories were simply too small for an individual to identify with in a meaningful sense.[3]

The theory that economic exchange results in numerous cultural connections was especially true for the sixteenth century. What one can accomplish these days through usual communication channels, such as telephone, telegram, checking account, and branch establishment, had then to be accomplished via large amounts of individual travel that required much time, money, and energy. This, in turn, led to many personal and intellectual ties among the merchants, intellectuals, artisans, and artists in various localities throughout Germany.[4] What has been referred to earlier in this work as more intensive communication within certain regions, refers primarily to this phenomenon.

The regio-centric orientation of the German southwest gave the border

---

[1] Karl S. Bader, "Die oberdeutsche Reichsstadt im Alten Reich," Esslinger Studien, XI (1965), 25-26.

[2] Erwin Hölzle, Beiwort zu: Der deutsche Südwesten am Ende des alten Reiches (Stuttgart: Kohlhammer, 1938), p. XX.

[3] Fritz Schnellbögl, "Die fränkischen Reichsstädte," Zeitschrift für bayerische Landesgeschichte, XXXI (1968), 458.

[4] Georg Simmel, "Soziologie des Raumes," Schmoller's Jahrbuch für Gesetzgebung, XXVII (1903), 64-65.

areas strong economic and cultural cohesion with neighboring areas.[1] This political and economic decentralization, together with the multitude of political affiliations, countered the trend toward a uniform religious movement. The ideal of a unified, centripetal territory of Swabia could not be realized for two reasons: (1) the opposition of the two external powers, Switzerland and the Hapsburgs and (2) the lack of a dominant economic and cultural center within Württemberg itself. The two outside powers strengthened the centrifugal forces at the southern and southwestern periphery, making the upper Rhine Plain and Upper Swabia (between Lake of Constance and upper Danube) almost independent entities from the remainder of the Southwest. The sparsely populated highlands of the Black Forest and the Swabian Alb reinforced the isolation of these regions from the larger and more densely-populated duchy of Württemberg. On the northern and eastern flanks of southwestern Germany, economic and cultural connections with central Germany were strong and manifold. Hence, the dependence of Nuremberg, Würzburg, Wertheim, and other Franconian cities on preachers and ideas from Wittenberg and Leipzig. The most important thoroughfares across the Black Forest and Swabian Alb were the Kinzig Valley and the road between Ulm and Geislingen, the latter located in the Neckar-Fils Valley. Through these gateways were channeled influences from Switzerland to the south, and Strasbourg to the west.

The Black Forest constituted an important dividing line between two different political regions.[2] The territory of Baden on its west-side was characterized by the lack of large imperial or territorial cities with a high degree of economic and cultural-ecclesiastic centrality. Economically, it depended on the metropolitan centers of Strasbourg, Basle, and Freiburg for the sale of its agricultural products. Strasbourg was to retain its role as dominant metropole of the Southwest well into the seventeenth century, when it was replaced by Ulm.[3]

Possibly no other city in Europe experienced a more peaceful religious change than Strasbourg. The manifold centrality of this metropole on the upper Rhine destined it to become a leading diffusion center for the Reformation move-

---

[1] Karl S. Bader, Der deutsche Südwesten in seiner territorialgeschichtlichen Entwicklung (Stuttgart: Koehler, 1950), p. 59.

[2] Bader, Der deutsche Südwesten, pp. 90-91; and Fritz Ernst, "Zur Geschichte Schwabens im späteren Mittelalter," in Festgabe für Karl Bohnenberger 75. Geburtstag, ed. by Hans Bihl (Tübingen: Mohr, 1938), p. 80.

[3] Karl S. Bader, "Die Reichsstädte des Schwäbischen Kreises," Ulm und Oberschwaben, XXXII (1951), 51.

ment. In addition to hosting a harmonious circle of superior preachers and a magistracy noted for its farsightedness in religious and political affairs, Strasbourg also enjoyed a great number of highly-educated humanists, an educated burgher class, a tradition of mystical piety, and free exchange of ideas through a network of traditional communication routes that converged on the city.

Judging by the volume of trade, no other German thoroughfare experienced traffic as intensive as the Rhine Valley. As a result, it served a culturally integrative function by channeling the <u>Devotia moderna</u>, a late medieval religious movement, to the south, and the Swiss-type Reformation northward to the Palatinate, Hesse, and the Netherlands. The Rhine <u>Graben</u> constituted a natural link between Switzerland and the lower Rhine, as well as between Upper Swabia and the cities of Basle and Strasbourg.

Whereas the Kinzig Valley was an important Black Forest thoroughfare, the Kraichgau between Odenwald and Black Forest formed the bridge between the upper Rhine-Palatinate and the Swabian-Franconian territories. A gateway since ancient times, its roads were used by German kings, emperors, and merchants who helped to establish trade connections to the outside (Fig. 14). It was here, moreover, that the Zwinglian and Lutheran brands of Protestantism clashed with each other.

The tribal boundary between Swabia and Franconia has often been equated with the division between the Zwinglian-oriented southern part and the Lutheran-oriented northern section of southwestern Germany[1] (see Fig. 15). This tribal-linguistic border, which divided the duchy of Württemberg into a larger Alemannic and a smaller Franconian portion, became blurred after the thirteenth century. Consequently, the southern part of Franconia, specifically the lower Neckar and Hohenlohe areas, had become separated from the Main Valley further north. Though both fell under Lutheran influence, the latter obtained its preachers directly from Wittenberg, while southern Franconia relied on graduates from Heidelberg. By the fifteenth century, a good part of the Franconian virtue of self-sufficiency had been undermined due to its increasing orientation and dependence on the Alemannic Württemberg core area between Stuttgart, Esslingen, and Marbach.

---

[1] Erik Wolf, "Die Sozialtheologie Zwinglis," Festschrift für Guido Kisch, Rechtshistorische Forschungen. No. 26 (Stuttgart: Kohlhammer, 1955), p. 168.

The Relationship between Size of Town and the Occurrence of Preachers

Since we lack a map of cities and towns in the sixteenth century, it was necessary to create one on the basis of various existing population figures[1] from that period. Total or partial censuses in this period were rare and not necessarily reliable. Nevertheless, the available figures allow us to visualize the approximate size and distribution of cities and towns. Even so, categorizing cities and towns on population figures alone does not tell the whole story. The size of an urban unit should be understood in terms of its functions, that is, a town or city that was dependent on traffic or communication was often more important than its population would have suggested.[2] Furthermore, we must differentiate between imperial and territorial cities and towns. The former were able to act in a more self-assured and decisive manner regarding the Roman-Catholic clergy. By today's standards, most of these cities and towns would appear to be large villages. It must be remembered, however, that in the sixteenth century, "dwarf towns" with populations between 500 and 1000 had the qualities of today's small towns with a population between 5000 and 20,000.

The cities and towns for which figures are readily available[3] can be divided into the following five categories: (1) metropoles with over 12,000 inhabitants; (2) medium cities with 5000 to 12,000 inhabitants; (3) large towns with 2500 to 5000 inhabitants; (4) rural small towns (Ackerbürgerstädte) with 1000 to 2500 inhabitants; and (5) dwarf towns (Zwergstädte) with less than 1000 inhabitants. Towns with less than 2500 residents, either territorial or imperial[4]

---

[1] The primary source were the Städtebücher, edited by Erich Keyser, for Württemberg, Baden, Hesse, and northern Bavaria; the following indices were listed and then converted into population figures by the author: (1) Herdstätte (household) = 6 inhabitants, (2) family = 6 inhabitants, (3) taxpayer = 5 inhabitants, (4) burgher = 5 inhabitants. The conversion figure was not constant, but depended on the state of the economy. Whereas in relatively prosperous times 1 burgher counted as 5 inhabitants (between 4.9 and 5.1), the rate dropped to 3.5-3.6 during depressions, such as the one following the Thirty Years' War. Personal information of Franz Quarthal and Professor Hans Jänichen, Tübingen.

[2] Albrecht Timm, "Die Stadt des Mittelalters und die moderne Stadt," Studium Generale, vol. XVI (1963), No. 9, p. 519.

[3] The Städtebuch for southern Bavaria had not been published by early 1973. Hence, the city map remained incomplete for the eastern rim area of southwestern Germany.

[4] Aalen, Buchau, Giengen, Bopfingen, Wangen, Buchhorn, Offenburg, Gengenbach, Zell-Harmersbach, Wimpfen, Pfullendorf, Weil der Stadt. These imperial towns, inhabited by peasants and wine growers retained the character

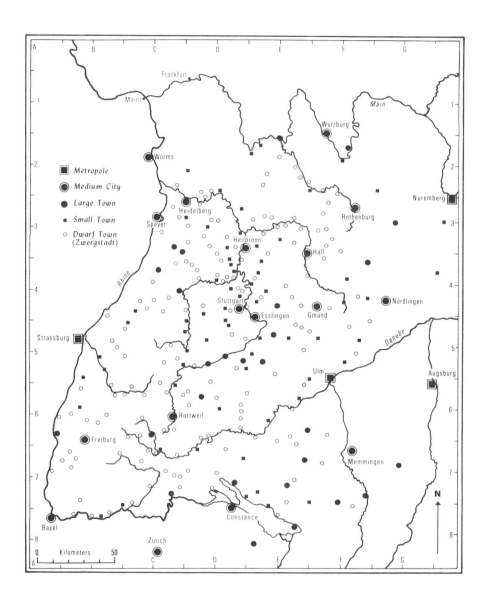

Fig. 12.--Cities and Towns of Southwest Germany at about 1550

Large, Small, and Dwarf Towns[1] Shown in Fig. 12

Aalen	F /4.2	Horrheim	D /3.8	Schorndorf	E /4.3
Ansbach	F9/2.9	Ingelfingen	E2/2.9	Schwabach	G1/2.9
Asperg	D3/4	Isny	F /7.4	Schwaigern	D2/3.3
Backnang	D8/4	Kaufbeuren	G /6.8	Sindelfingen	D1/4.5
Baden-Baden	B7/4.3	Kempten	F4/7.3	Sinsheim	C9/3
B. Liebenzell	C5/4.3	Kenzingen	A8/5.9	Steinbach	B6/4.5
Bad Waldsee	E4/6.7	Kirchheim/T.	D9/4.7	St. Gallen	D6/8
Balingen	C8/5.7	Kitzingen	F1/1.7	Sulz	C3/5.5
Beilstein	D7/3.5	Künzelsau	E3/3	Tauberbischofs-	
Besigheim	D3/3.7	Lahr	A9/5.4	heim	E2/2
Biberach	E5/6.2	Lauda	E3/2.2	Thayngen	C4/7.2
Bietigheim	D3/3.8	Langenau	F /5.1	Trossingen	C3/6.2
Binsdorf	C5/5.6	Leonberg	D1/4.3	Tübingen	D1/5.1
Blaubeuren	E5/5.4	Lichtenau	B3/4.4	Tuttlingen	C7/6.6
Böblingen	D2/4.6	Lindau	F3/7.8	Überlingen	D3/7.1
Bönnigheim	D2/3.6	Lorch	E3/4.2	Unteröwisheim	C8/3.2
Bopfingen	F4/4.1	Löwenstein	D8/3.4	Urach	D7/5.2
Brackenheim	D2/3.5	Marbach	D6/3.9	Vaihingen/E.	D /3.8
Breisach	A4/6.3	Markdorf	D7/7.2	Villingen	C /6.3
Bretten	C5/3.5	Markgröningen	D2/4	Waiblingen	D6/4.2
Bruchsal	C3/3.3	Meersburg	D5/7.3	Waldshut	B5/7.5
Bühl	B5/4.6	Mergentheim	E5/2.4	Wangen	E5/7.4
Buchhorn	D8/7.5	Metzingen	D6/5	Weil der Stadt	C8/4.4
Calw	C5/4.5	Miltenberg	D6/1.8	Weilheim	E2/4.8
Dinkelsbühl	F4/3.5	Mosbach	D4/2.7	Weinheim	C5/2.1
Donaueschingen	C1/6.6	Munderkingen	E3/5.9	Weinsberg	D6/3.3
Durlach	C1/3.7	Nagold	C5/5	Weissenburg	G5/3.7
Ebingen	D1/5.9	Neckarbi-		Wertheim	E /1.6
Ehingen	E4/5.8	schofsheim	D/2.9	Wiesloch	C5/2.9
Ellwangen	F1/3.8	Neckarstei-		Wildbad	C1/4.4
Eppingen	C9/3.3	nach	C8/2.5	Wildberg	C5/4.7
Ettlingen	C /3.8	Neudenau	D6/2.9	Wimpfen	D4/3.1
Feuchtwangen	F4/3.2	Neuenbürg	C2/4.1	Windsheim	F6/2.4
Freudenberg	D7/1.7	Neuenburg	A4/6.9		
Fürfeld	D2/3.2	Neuenstadt	D7/3.1		
Geislingen	E6/4.8	Nürtingen	D7/4.8		
Gengenbach	B2/5.3	Oberriexingen	D1/3.9		
Gerabronn	E7/3	Ochsenfurt	F /1.9		
Gernsbach	B9/4.3	Offenburg	B1/5.1		
Giengen	F3/4.8	Öhringen	E /3.2		
Gönningen	D3/5.2	Pforzheim	C4/4		
Göppingen	E3/4.6	Pfullendorf	D4/6.7		
Gross-Bottwar	D7/3.7	Pfullingen	D5/5.3		
Gross-Garlach	D3/3.3	Ravensburg	E1/7.1		
Gr. Sachsenheim	D2/3.8	Reutlingen	D4/5.1		
Haiterbach	C4/5.1	Riedlingen	D8/6.1		
Heidelsheim	C5/3.4	Rottenburg	C9/5.2		
Herrenberg	C8/4.8	Saulgau	E /6.4		
Hirschhorn	C9/2.4	Schaffhausen	C3/7.3		
Horb	C5/5.2	Schömberg	C6/5.9		

[1] Only those dwarf towns which have a preacher were listed.

were considered to be of far less economic importance than the first three categories (Fig. 12).

Matching the population of cities and towns with the date of their first evangelical sermon gives us a glimpse of the temporal and spatial diffusion of the Reformation in southwestern Germany. Since the year 1518 can be considered as the earliest possible date for delivering a sermon, the average time-span between that date and the actual year of the first sermon could be computed for the various categories of cities and towns. They are arranged in the following order:

(1) Metropoles: Strasbourg, 1520; Ulm, 1521; Augsburg, 1518; Nuremberg, 1518. This gives us an average of $1\frac{1}{4}$ years.

(2) Medium-sized cities: Freiburg, 1521; Esslingen, 1522; Gmünd, 1523; Zurich, 1519; Constance, 1519; Memmingen, 1523; Hall, 1523; Heilbronn, 1524; Rothenburg, 1521; Würzburg, 1519; Heidelberg, 1518; Speyer, 1524; Rottweil, 1524; Basel, 1520 (?); Stuttgart, 1520; Nördlingen, 1521. This amounts to an average of not quite 3 years and 4 months.

(3) Larger towns: Balingen, 1523; Tübingen, 1525; Windsheim, 1522; Biberach, 1523; Kitzingen, 1522; Durlach, 1532; Bruchsal, 1522; Heidelsheim, 1525; Pforzheim, 1519; Rottenburg, 1523; Urach, 1522; Kirchheim/Teck, 1530; Villingen, 1525; Überlingen, 1525; Lindau, 1522; Isny, 1522; Leutkirch, 1530; Kempten, 1524; Kaufbeuren, 1521; Dinkelsbühl, 1523; Wertheim, 1522; Schorndorf, 1534; Bad Waldsee, 1530. The average amount of time for the introduction of the evangelical sermon was 6 years, 1 month.

(4) Small towns: Lahr, 1524; Offenburg, 1525; Baden-Baden, 1524; Brackenheim, 1520; Beilstein, 1524; Öhringen, 1534; Wangen, 1530 (?); Überlingen, 1529; Riedlingen, 1520; Ehingen, 1524; Vihingen, 1534; Weinsberg, 1520; Weil der Stadt, 1522; Nürtingen, 1534; Ebingen, 1525; Marbach, 1534; Bopfingen, 1522; Kenzingen, 1522; Gengenbach, 1523; Rastatt, 1534; Tuttlingen, 1534; Mosbach, 1520; Eberbach, 1534; Hockenheim, 1534; Aalen, 1526; Mergentheim, 1525 (?); Königsheim, 1534; Lauda, 1524; Freudenberg, 1527; Miltenberg, 1523 (?); Schwai-

---

of a territorial town. In spite of their guild constitution, they were unable to work it to their advantage. Aalen and Bopfingen were without guilds. Siegmund Keller, "Der Adelsstand des süddeutschen Patriziates," in Festschrift Otto Gierke zum 70. Geburtstag (Weimar: Böhlau, 1911), p. 750.

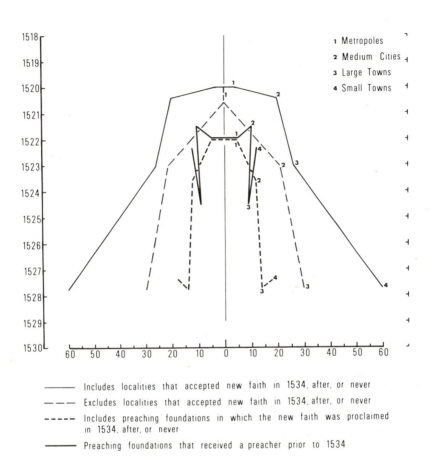

Fig. 13.--Relationship between the Size and Number of Cities and Towns and the Year of Adopting a Preacher of the New Faith.

gern, 1525; Weilheim, 1534; Herrenberg, 1534; Nagold, 1534; Wiesbach, 1534; Leonberg, 1534; Bönnigheim, 1525; Backnang, 1534; Bottwar, 1525; Donaueschingen, 1534; Geislingen, 1527; Buchhorn, 1534; Meersburg, 1534; Waldshut, 1522; Laufburg, 1534; Besigheim, 1524; Pfullendorf, 1534; Markdorf, 1534; Pfullingen, 1534; Metzingen, 1534; Sinsheim, 1534; Wiesloch, 1534; Laupheim, 1534; Wimpfen, 1523; Neckarsulm, 1534; Wildberg, 1534. The average space of time that passed before the introduction of the first evangelical sermon in small towns was 10 years and 7 months.

These data support the following fact: The higher the degree of urbanization, the more marked was the interest of the public in the innovation (Fig. 13). The environment within larger cities and towns proved to be the most unstable for the old faith. Here, the ruling body or magistracy, was obliged to act according to the wishes and desires of the burghers, who showed a strong predilection for the new faith.[1] In general, the grievances against the traditional ecclesiastical establishment were more manifest here than in rural areas. The cities were the stronghold of the humanist movement, which was responsible for the burghers' quest for autonomy against a fully organized church.

---

[1] An exception was Freiburg, a city that had not experienced a "guild revolution." Its magistracy was therefore particularly harsh in suppressing the evangelical movement in the city and in neighboring towns. In other cities, the guild revolution provided the constitutional framework in which its political and religious conflicts were carried out.

CHAPTER II

THE REFORMATION AND CONCOMITANT SOCIAL

AND RELIGIOUS MOVEMENTS

From the point of view of sociology, the religious realm performs a double function: On the one hand, it provides the major basis for legitimating an existent social order; on the other hand, it may be a major source of inspiration for critical or utopian thoughts.[1] The evangelical movement, by rejecting the established tradition or ideology of the Roman-Catholic Church, and substituting for it a scheme for an ideal religious order, clearly belongs in the second category. It was both a protest as well as a religious phenomenon, related to each other in the form of two concentric circles, the larger representing the social protest movement, which to some extent depended on the inner core of religiously-motivated persons.[2] Religious reforms, originally meant to stand separate from secular affairs, injected elements of unrest into the political and social order of cities[3] and rural territories,[4] thus reinforcing the religious

---

[1]Tiryakian, "A Model of Societal Change and its Lead Indicator," p. 82.

[2]Rösler, Geschichte und Strukturen der evangelischen Bewegung im Bistum Freising 1520-1571, p. 42.

[3]Hans Mauersberg, Wirtschafts und Sozialgeschichte Zentraleuropäischer Städte in neuerer Zeit (Göttingen: Vandenhoeck & Ruprecht, 1960), pp. 104-13.

[4]In the words of Adolf Waas: "The religious-social unrest among peasants as well as lower urban strata in 1525 must be regarded as the most important factor leading to the Peasants' War. It was the deciding impulse that accelerated and reinforced all the other latently existing causes of the movement." Adolf Waas, Die Bauern im Kampf um Gerechtigkeit 1300-1525 (München: Callway, 1964), p. 62. The Klettgau illustrates this point. At first, its peasants, content with their political and economic situation, rejected requests by rebelling neighbors to join them. This changed when preachers entered the area, accusing the political authorities of willfully denying the gospel to the common man. The receptivity for the evangelical creed was increased by telling the populace how unlawful and un-Christian traditional property rights with levies and services to be rendered, were. Josef Bader, "Aus der Geschichte des Pfarrdorfes Griessen im Klettgau," Freiburger Diözesan-Archiv, IV (1869), 240.

movement. It can be said, therefore, that the success of the Protestant movement was due to the interpenetration of religion, politics, and society. City magistracies and territorial governments feared Luther's and, particularly, Karlstadt's and Münzer's ideas for their disruptive effect on the existing social and political order. These radical reformers created a negative image of the religious movement among such city governments and rulers as those of Schwäbisch Gmünd, Rothenburg/T., Windsheim, Ulm, Augsburg, Nuremberg, and Wertheim. For example, in Ulm and Nuremberg, the mere gathering of crowds around a preacher outside the church elicited fears that the public order might be disturbed. Hence, the city governments tried to get rid of their itinerant preachers.

Popular movements, which included the various peasant and religious uprisings, were dependent on three variables: (1) their leading personalities or opinion leaders, (2) their background relating to existing social and economic conditions, and (3) patterns of organization of its adherents that lead to the realization of the movement. All three factors were inter-twined, the latter two depending heavily on the influence of the opinion leaders. The chances for the realization of a reformed church were heightened by socio-economic conflict situations that were supra-individual, structurally-dependent events of special importance. In a conflict that comprises two or more parties, an individual may be involved against his will and can disentangle himself only at the cost of alienating himself from the reference group that shares his social and economic status.[1] The anticipation of each other's actions, shared by members of a special group, can and did extend beyond the concrete situation, to symbolic systems and beliefs.[2]

In the following pages, a comparison is drawn between localities that experienced unrest among their rural folk and the acceptance of the evangelical creed in the form of sermons. A complete or even high correlation is impossible owing to the political and/or personal factors that enter as a filtering system. The latter often proved to be decisive in the toleration or non-toleration of the new faith and were of increasing importance as the Reformation movement proceeded. In addition, a preacher had to be found who was ready to advocate the new line. Frequently the movement reached a beginning stage, but was

---

[1] Ernst Buchholz, Ideologie und latenter sozialer Konflikt, Göttinger Abhandlungen zur Soziologie und ihre Grenzgebiete (Stuttgart: Enke, 1968), p. 12.

[2] The Social Sciences in Historical Study, Social Science Research Council, Bulletin No. 64 (New York, 1954), p. 54.

short of a preacher who could articulate and realize the reformation ideals.

The decades prior to the Reformation were marked by a series of great social upheavals in the rural sector. It began in the Alsatian region in 1493 and ended in 1525 with the Peasants' War. The points made in the Reformatio Sigismundi are so similar to the demands made by the peasants in 1525, that the former has been called the "trumpet of the Peasants' War." Protest against the increasing exactions under which the peasants suffered is common to both the Reformatio and the famous Twelve Articles of 1525; hence the Reformatio was itself re-published in a number of editions between 1521 and 1525.[1]

Actually, the term Peasants War is a misnomer, since this revolutionary movement swayed the smaller cities and towns as well. The further north the movement advanced, the more it became an urban-based phenomenon. Some authors have therefore suggested the term "rebellion" (Empörung) as being more appropriate.[2]

An immediate result of attempts to encourage religious reform was that an increasing number of peasants rejected tithes to their landlords. Furthermore, individual parishes demanded the right to select their own preachers. This had its strongest appeal in Switzerland, from where the preachers travelled across the border into Germany in order to serve as instigators for the various local rebellions. The demand for "divine righteousness" formed the basis for a vast movement, ranging from Switzerland and Austria to Hesse and Saxony. It united the numerous localized revolts into a larger supra-regional one. Hence, for the first and only time a major religious innovation was accepted in rural areas.[3]

The Peasants' War

The Hegau and Klettgau provinces that border Switzerland to the north were the first areas in Germany to feel the effect of the Swiss peasant unrest.

---

[1] William R. Hitchcock, The Background of the Knights' Revolt 1522-1523, University of California Publications in History, No. 61 (Berkeley and Los Angeles, 1958), p. 100.

[2] Eberhard Mayer, Die rechtliche Behandlung der Empörer von 1525 im Herzogtum Württemberg, Schriften zur Kirchen- und Rechtsgeschichte, No. 3 (Tübingen: Osiander, 1957), p. 2.

[3] Günter Franz, "Das Verhältnis von Stadt und Land zwischen Bauernkrieg und Bauernbefreiung," Studium Generale, XVI (1963), 558.

Here the "poor man" or peasant[1] examined his social standing much more critically than in other German provinces, for he enjoyed direct contact with his associates south of the Rhine who attempted to free themselves from their noble overlords. What is more, the population of the Klettgau enjoyed the civic rights that were so relished by the residents in Zurich. These were the very terms which the local count had to accept after he lost several small wars with Switzerland.[2]

The peasant unrest in this border area began in May 1524, although requests for religious changes were not voiced until the end of that year. After the Waldshut preacher, Dr. Balthasar Hubmayer had begun to preach the gospel, Zwinglian preachers arrived in the Duchy of Sulz from Schaffhausen and Zurich. By January 1525, the Klettgau peasants denied the small tithe to their abbot, who had refused to honor their demand for self-appointment of a Protestant preacher.[3] Harsh counter-Reformation measures initiated by local territorial leaders and the Hapsburg met with strong resistance from the townspeople, especially those of Waldshut who, with the support of Zurich, played the leading role. The entire area immediately north of the Swiss-German border was oriented towards, and existed in a socio-economic equilibrium with Zurich, which served as an economic, cultural, and religious central place.[4]

The village of Griessen was the focal point for the uprising in the Klettgau. After the Catholic priest fled in the wake of the rebellion, the parish frequently sent a representative to the chapter in St. Blasien, requesting a preacher of the new faith; but to no avail. Finally, after searching on its own, the community found Johann Rebmann, a Swiss from Vigoltingen in the canton of Thurgau. Educated in Waldshut and Strasbourg, he preached the new gospel in Kleeburg near Weissenburg, Alsace. On the request of Zurich, this Zwinglian preacher came to Griessen, where he became one of the most arduous supporters of ecclesiastical and political change. The movement was suppressed, but when the count again raised the level of contributions a century later, the peasants once more

---

[1] Armer Mann or Arme Leute was a medieval expression that applied to the rural population as a whole, no matter whether rich or poor.

[2] Bader, "Aus der Geschichte des Pfarrdorfes Griessen im Klettgau," pp. 230-31.

[3] Günter Franz, Der Deutsche Bauernkrieg (4th ed.; Darmstadt: Gentner, 1958), p. 110.

[4] Peter Schöller, "Der Markt als Zentralisationsphänomen," Westfälische Forschungen, XV (1962), 91.

asked Zurich for help. Bergöschingen, likewise, acquired a Zwinglian preacher in the early part of 1525, which means that the movement would have quickly covered the entire Klettgau, had it not been suppressed by its ruler.

The area that displayed strongest the interrelation between religious change and peasant unrest was Upper Swabia and the Allgäu. The reformers in Memmingen and Kempten bolstered the peasants' demands for social and political conditions based on the new teaching. Beginning in 1523, the imperial city of Memmingen with its rebellious and outspoken foundation preacher Dr. Christoph Schappeler became the center for the reform movement. Dr. Schappeler can be credited for the diffusion of revolutionary ideas that were readily accepted by a sizable group of rural clergymen and peasants north and south of Memmingen.

The heavy concentration of peasant preachers in the Kempten area can be explained by the harsh demands of its prince abbot, who held many peasants in serfdom. They tried all possible avenues to escape his severe rule. Peasant revolts against both church and nobility in 1492 and 1525, respectively, were caused by oppressive measures and foreign influence on the peasants. In 1523, peasants in 17 parishes[1] refused to pledge their allegiance to Abbot Sebastian of Breitenstein unless he heard their complaints. The abbot flatly refused their request. St. Mang parish in Kempten was the first to reject the oath of allegiance under Kaspar Heelin, its Prädikant (foundation preacher). Matthias Waibel, a peasant's son who had studied at the University of Vienna became parson at Kempten's St. Lorenz.[2] When the Swabian Federation, a counter-revolutionary organization, accused Waibel of proclaiming the evangelical faith, Kaspar Heelin offered him refuge at St. Mang. Someone else in this parish, Jörg Knopf, was to assume the leadership of the peasants as they assembled in his village at Leubas. As with most peasant leaders, he lacked a strong religious conviction. The peasant preachers were Christian Wanner in Haldenzwang, Walter Schwarz in Martinszell (the birthplace of Matthias Waibel), Andreas Stromair and Hans Ul in Oberdorf, Mang Batzer in Buchenried, and Florian in Aichstetten; Matthias Röt in Memhölz, Hans Haering in Legau, Hans Unsinn in Oberthingau, Hans Hafenmayer and Veit Riedle in Günzburg. The majority sought refuge in Appenzell and Rorschach (Switzerland) after the peasants'

---

[1] By 1525, not fewer than 26 villages filed a complaint. See Otto Erhard, Der Bauernkrieg in der gefürsteten Grafschaft Kempten (Kempten: Kösel'schen, 1908), p. 13.

[2] Otto Erhard, Matthias Waibel (Kempten: Kösel'schen, 1925), pp. 1-14.

defeat.[1]

Suppression of the Kempten abbey peasants did not mean an end to the demands for social and religious renewal. In June 1526, 800 peasants assembled in an open field near Wiggensbach around the lay preacher Hans Häberlin, as well as outside the villages of Grönenbach and Altusried. Häberlin strongly polemicized against Lutheran "monks, sophists, and Antichrists" who argued that Christ is "inside or accompanying" the Lord's Supper.[2]

Further north, the peasants of the imperial city of Ulm remained quiet, although their political status was not unlike that of many subjects who suffered under a territorial ruler. The town of Leipheim had since the summer 1523 Hans Jakob Wehe (a cousin of Eberlin of Günzburg) as a gospel preacher. The city of Ulm, on request by the Swabian Federation, demanded his removal. That provided the spark for armed opposition, and in late February and early March 1525, the burghers of Leipheim joined the rebellious peasants in the country.

The peasant unrest north of the Swabian Alb had its origin in the imperial city of Rothenburg/Tauber. There, the Reformation was introduced in 1521 with social-revolutionary programs in which the burghers tried to recapture their former political influence. In this case, burghers and peasants were united by a similar objective. The foundation preacher of Rothenburg, Dr. Johann Teuschlein, was the rallying point for their dissatisfactions.[3] The two parsons in Lenzenbronn near Rothenburg, Lienhard Deuner and Hans Hollenpach, were the peasant leaders who served as links between the revolting factions in the city and country. They opposed the tax system of the city, which was already the target of attack by Dr. Teuschlein.[4] The craftsmen who were responsible for the urban opposition likewise demanded a decrease in taxes. Following the defeat of the peasants, 200 Rothenburgers successfully petitioned the old magistracy to reinstate the mass.

---

[1] Friedrich Dobel, Das Reformationswerk zu Memmingen unter dem Drucke des Schwäbischen Bundes 1525-1529 (Augsburg: Lampart, 1877), p. 23.

[2] Erhard, Der Bauernkrieg, p. 91, footnote 1.

[3] Franz, Der Deutsche Bauernkrieg, p. 178. Teuschlein advocated a united front with the peasants, provided that the latter associate their actions with the evangelical faith. See Theodor Kolde, D. Johann Teuschlein und der erste Reformationsversuch in Rothenburg o.d.T. (Erlangen: Böhme, 1901), p. 34.

[4] Franz, Der Deutsche Bauernkrieg, pp. 179-80.

Rothenburg also influenced the Tauber Valley below the city. Although they did not rely on the Twelve Articles, the peasants there demanded first and foremost a preacher of the gospel who would speak every day to the migrating crowd of revolting peasants. One of their most important contributors was a Florian Geyer, who skillfully negotiated between this group and various cities and towns that could be regarded as allies. As in the case of the peasants, he regarded the Bible and Luther's teaching as the foundation for his political aspirations.[1]

Unlike Rothenburg/Tauber, Schwäbisch-Hall did not manifest any overwhelming burgher unrest. The city asked its craftsmen to remember their earlier oath not to join their peasant companions in the countryside. After the peasants and their religious leaders to the north, south, and southeast of the city[2] had been encouraged by Johannes Walz, schoolmaster and supervisor of the local Franciscan monastery, to oppose the tithe, Johannes Brenz and his colleague Johann Herold at Reinsberg took a strong stand against the peasants' demands.[3]

However, the towns of Gaildorf and Ellwangen did join the rebelling peasants. On April 17, 1525, the peasants of Limpurg county met in Gaildorf. They had ties to the rebels in Franconia and Württemberg. In neighboring Ellwangen, the early Reformation movement that began in 1521 was purely religious in nature and limited to the town. In the wake of the civil disorders ensuing from the peasant uprising, its leaders, Georg Mundtpach and Dr. Johann Kress (city parson and foundation preacher, respectively), completely abandoned the old liturgy. It was primarily Mundtpach who led the drive against the established church and, finally, refused to retract, whereas Kress eventually repudiated the position he had taken against the Catholic Church. The magistracy of the imperial city of Dinkelsbühl under pressure from sympathetic burghers likewise opened its gates to the rebelling peasants, as did Nördlingen.[4]

In Württemberg, the expedition of Duke Ulrich in spring 1525 further

---

[1] Ibid., p. 187.

[2] Orlach, Tauberzell, Bühlertann, and Welzheim. See Gustav Bossert, "Drei Haller Biographien," Württembergisch-Franken, VII (1900), 75-76.

[3] Julius Gmelin, Hällische Geschichte (Schwäbisch Hall: Staib, 1896), pp. 713 ff.

[4] Beschreibung des Oberamts Ellwangen (Stuttgart: Kohlhammer, 1886), pp. 496-97.

weakened the already fragile hold of the Austrian government over this territory. Everywhere, the rebellion gained strength. In the small town of Bottwar, 200 inhabitants banded together under Matern Feuerbacher and joined another gang at Pfaffenhofen on the left side of the Neckar. On April 25, the peasants occupied Stuttgart and forced the government to relocate in Tübingen. By this time, the entire country was under the control of the peasant renegades.[1]

In northern Baden, Brurain peasants occupied the bishop's residence and demanded the "Gospel and fairness." They were joined by the residents of Bruchsal as well as the bishop's soldiers. When Margrave Phillip promised to give in to their demands, the peasants disbanded.[2]

Regardless of whether or not the reformatory slogans provide a basis for socio-political unrest, suppression by local authorities clearly had a negative effect on the religious movement. As the social, economic, and political demands were left hanging in the air, so the religiously-oriented demands remained unfulfilled. This in part explains the loss of momentum experienced by the religious movement after the Peasants' War. Whereas most villages and towns lost their evangelical preacher, imperial towns witnessed some kind of co-existence of the old and the new faith. Neither possessed or was allowed enough political and social sustenance or potential to replace the other. But the war did not show the same effect everywhere. Whereas the area that was controlled by the Swabian Federation experienced a temporary curtailment or reversal of the Protestant movement, the Peasants' War prepared a number of communities in Baden for Lutheran preachers. Here, in the words of Kattermann, "the Peasants' War reinforced the courage of the population for the new confession."[3]

The unrest at Bauerbach and the Speyer prebendary was closely connected with the spread of Lutheran ideas. But in spite of the strong measures taken by the bishop and chapter, the new faith gained adherents in rural districts as well. In fact, the Lutheran movement gained ground particularly in areas where the revolt was strong.[4]

---

[1] Franz, Der Deutsche Bauernkrieg, pp. 216-21.

[2] Ibid., pp. 222-23.

[3] Kattermann, Die Kirchenpolitik Markgraf Phillip I, p. 41.

[4] Claus Peter Clasen, Die Wiedertäufer im Herzogtum Württemberg und in benachbarten Herrschaften, Veröffentlichungen der Kommission für Geschichtliche Landeskunde, Serie B, No. 32 (Stuttgart: Kohlhammer, 1965), p. 61.

It was at this time that the Reformation became separated from the social trends of the time, with a re-emphasis on the purely religious core of the movement. In 1525, the lay movement, an important propagator of the evangelical movement, experienced a significant decline and was replaced by the clerical movement. The visitations that followed set the tone for the increasing institutionalization of the Lutheran creed.[1]

The effect of Luther's condescending remarks about the revolting peasants has been greatly overestimated. His ideas were too well entrenched by then to allow a slackening of interest. The physical and social distance between Luther and the majority of peasant and townspeople was simply too great to have any pronounced effect. The appearance of preachers in Bietigheim, Reichenbach, and Wiggensbach in 1526, in Jöhlingen, Asperg, Sernatingen, Saulgau, and the umland surrounding Esslingen in 1527, and Schlath in 1528, suggests a continued, yet politically suppressed enthusiasm for the evangelical message. Not until 1528-1529 did the movement pick up steam again, though by this time through the effort of foreign preachers instead of its own local preachers.

The Relevance of Marxist Theory

Marxist theoreticians and historians contend that when the socio-economic status of the proletariat is low compared to that of the ruling bourgeoisie, the revolutionary potential is great enough for the successful launching of a revolution. An attempt was made in the course of this study to explore whether or not there was any correlation between areas with an above-average percentage of poor people and their readiness to accept a new religious ideology as a symbol of social protest. It should be made clear that such a connection could only have existed prior to the end of the Peasants' War.

To test this thesis, it was essential to find data about economic well-being for a larger political entity. The tax lists (Herdsteuerlisten) on homes and other possessions for the year 1525, tabulated for the duchy of Württemberg, and preserved in the Hauptstaatsarchiv in Stuttgart, is an ideal source for this kind of investigation. The author computed the percentage of non-taxpayers and compared it to the number of persons who had their own home, which included the vast majority of taxpayers; it was also compared with those who did not own their homes but had to pay taxes on other possessions. This computation was

---

[1] Hans Liermann, "Laizismus und Klerikalismus des evangelischen Kirchenrechts," Zeitschrift der Savigny-Stiftung für Rechtsgeschichte, Kan. Abt., LXX (1953), 6-7.

done for 27 prefectures (Ämter),[1] and the same procedure was then applied to 31 towns, the administrative seats of these and other prefectures, in order to ascertain their economic status or economic centrality (Tables 1 and 2).

A brief description of the relationship between the Amtsstadt or administrative town, and Amt or prefecture is necessary. The towns in their administrative functions were all similar inasmuch as they performed a surplus of duties. They usually enjoyed strict control over the surrounding rural districts. The wide-spread social unrest of the time, beginning with the Bundschuh[2] and ending with the Peasants' War, signalled the common peoples' demand for political equality with the towns. In most prefectures, especially in wine-growing regions, the villagers in toto paid more taxes than the townspeople. The latter seldom provided more than one-third of the total tax revenue and in some cases much less. Naturally, this situation burdened the peasants with a heavy tax load. Nevertheless, in all the prefectures, differences in the economic and social structure of towns and villages were slowly fading. In many towns, agriculture and viticulture played an important role, while many villages in densely-populated wine-producing regions were becoming townlike in the number of inhabitants and their functions.[3] Also, rural areas attracted more and more of the manufacturing done in cities.

An analysis of the data of prefectures shows that only one of them, Urach, with an above-average percentage of "have-nots" had a preacher. Even the prefectures of Calw and Wildberg, with more than twice the average percentage of non-taxpayers, including a large proportion of people employed in textile manufacturing, never hosted a Lutheran preacher. Ironically, textile workers have been thought to be especially susceptible to the new faith.[4] Waiblingen, Brackenheim, Böblingen, Urach, Tübingen, Besigheim, and Neuenbürg were the only

---

[1] The tax lists for several prefectures, notably Stuttgart, fell victim to World War II bombing raids.

[2] Name of a peasants' confederation.

[3] Walter Grube, "Stadt und Amt im Altwürttemberg," Stuttgarter Zeitung, June 29, 1972, p. 35.

[4] Anton Störmann considered the textile workers as the principal apostles of Lutheranism, owing to their economic malaise. See: Die städtischen Gravamina gegen den Klerus am Ausgange des Mittelalters und in der Reformationszeit, Reformationsgeschichtliche Studien und Texte, No. 24-26 (Münster: Aschendorffsche, 1916), p. 148; also, Kurt Kaser, "Zur politischen und sozialen Bewegung im deutschen Bürgertum des 15. und 16. Jahrhunderts," Deutsche Geschichtsblätter, III (1901), 56.

Amtstädte where sermons could be heard. In Waiblingen and Brackenheim, the Lutheran doctrine was interpreted by the local foundation preacher and in Urach and Tübingen by Carthusian and Augustinian monks. Peasants in and around Urach zealously supported the rural disturbances in 1514 and 1525. An evangelical preacher, whose name remains unknown, appeared in Urach in 1525, but was immediately executed by the bailiff.[1]

The large number of non-taxpayers in Neuenbürg seems to have no connection to the appearance of a preacher there. Since this town was in the possession of Franz of Sickingen, an ardent supporter of the Lutheran movement, it is safe to assume that he ordered the local preacher. His death in 1523 signalled the formal end of a noticeable movement in this town, although the Austrian government hesitated to punish adherents of the new religion for fear of greater popular unrest.[2]

An analysis of the towns on the basis of their average revenue per taxpayer, an indicator of their relative prosperity, and an examination of their propensity towards the new faith offers a different picture than what Marxist scholars would suggest. Out of 31 administrative towns, 16 had an individual property base above the median (75 guilders); and not less than 12 of these, or 75 percent, hosted a preacher. On the other hand, of the 15 remaining towns with an individual property base <u>below</u> the median, only 6, or 40 percent, could boast a preacher.

If wealth could be said to favor the new teaching, this ratio of almost two to one between the wealthy and the less wealthy towns in this sample test might well prove the point. In other words, wealth rather than poverty seems to have been the key indicator of the success of the new religious movement.

A good example of this hypothesis is offered by the village of Überkingen, a prosperous spa west of Ulm. In a letter to the mayor and magistrate of Ulm, Ambrosius Blarer mentions the eagerness of its citizens to adopt the new faith.[3] Many of them regularly attended Paul Beck's sermons in neighboring Geislingen.[4] No doubt, vacationers to Überkingen also had their share in familiarizing its

---

[1] D. Leube, ed., Im Zeichen von Sankt Christoph (Urach: Benz, 1928), p. 83.

[2] Beschreibung des Oberamts Neuenbürg (Stuttgart: Aue, 1860), pp. 92-93.

[3] Schiess, ed., Briefwechsel der Brüder Ambrosius und Thomas Blaurer, I, 258.

[4] Georg Burkhardt, Geschichte der Stadt Geislingen an der Steige (Konstanz: Thorbecke, 1963), I, 180.

TABLE 1

LIST OF PROPERTY TAX BY PREFECTURES[1]

Prefecture (Ämter)	Number of Taxpayers	Number of Non-taxpayers	Comparative Percentage of Non-taxpayers to Taxpayers (%)
Backnang	751	73	9.7
Balingen	762	58	7.6
Beilstein	457	16	3.5
Besigheim	374	28	7.5
Bietigheim	460	29	6.3
Böblingen	1236	102	8.3
Bottwar	308	12	3.9
Brackenheim	767	64	8.5
Calw	734	139	19.0
Cannstadt	1299	128	9.9
Dornstetten	348	51	14.7
Güglingen	540	19	3.5
Herrenberg	953	86	9.0
Kirchheim	1756	88	5.0
Lauffen	554	39	7.0
Leonberg	1200	50	4.2
Marbach	974	26	2.6
Maulbronn	1383	64	4.6
Nagold	417	45	10.8
Neuenbürg	652	55	8.4
Neuenstadt	423	23	5.7
Rosenfeld	384	50	13.0
Schorndorf	2502	140	5.9
Tübingen	1694	168	9.9
Urach	1150	159	13.8
Waiblingen	768	62	8.1
Wildberg	444	88	19.8
Total	23280 (100%)	1862 vs.	8.0 8.0%

The Median Percentage of the Non-taxpayers compared to Taxpayers is 8.0%.

---

[1] According to Herdsteuerlisten for 1525. Haupstaatsarchiv Stuttgart, Bestand A54a, Lists 19-50.

TABLE 2

LIST OF PROPERTY TAX BY ADMINISTRATIVE TOWN (AMTSSTADT)

Town (Amtstadt)	Number of Taxpayers	Number of Non-taxpayers	Comparative Percentage of Non-taxpayers to Taxpayers (%)	Average Amount of Tax per Individual (in guilders)
Backnang	219	21	9.6	52
Balingen	199	11	5.5	70
Beilstein	94	4	4.2	55
Besigheim	199	22	11.1	84
Bietigheim	206	17	8.3	140
Blaubeuren	178	25	14.2	75
Böblingen	141	14	9.9	70
Bottwar	192	6	3.1	79
Brackenheim	203	18	8.9	86
Calw	288	91	31.6	57
Cannstadt	388	57	14.6	98
Dornstetten	107	23	21.5	73
Ebingen	175	16	9.1	65
Güglingen	109	1	0.9	91
Herrenberg	283	32	11.3	90
Kirchheim	458	22	4.8	78
Lauffen	81	5	6.2	89
Leonberg	147	16	10.9	84
Marbach	262	14	5.3	14
Maulbronn	96	10	10.4	33
Nagold	178	25	14.0	69
Neuenbürg	51	20	39.2	87
Neuenstadt	146	14	9.6	80
Rosenfeld	77	11	14.3	37
Schorndorf	412	43	10.4	126
Tübingen	579	84	14.5	92
Urach	280	65	23.2	76
Waiblingen	258	30	11.6	140
Weinsberg	250	14	5.6	24
Wildbad	95	11	11.6	51
Wildberg	142	45	31.7	62
Total	7363	787	10.7	75

The Median Comparative Percentage of the Non-taxpayers to Taxpayers is 10.7%.

---

[1] According to Herdsteuerlisten for 1525; Hauptstaatsarchiv Stuttgart, Bestand A54a, Lists 19-50.

residents with the new ideology. Likewise, many of the prefectures, as well as their main towns, also enjoyed considerable prosperity through viticulture and wine-trade with cities and towns in Bavaria.

It may well be impossible to judge the stability or instability of any given society on the basis of socio-economic or "objective" conditions alone, which tend to mask more "subjective" or symbolic factors. Nevertheless, the socio-economic interpretation can be regarded as only <u>one</u> explanation rather than <u>the</u> explanation of the early diffusion of the Lutheran doctrine.

CHAPTER III

THE IMPACT OF HIGHER EDUCATION AND COMMUNICATION

ON THE REFORMATION

Universities and the Reformation

In view of the previous discussion on the role of institutions, Gustav Benrath's statement that "the Reformation of the church followed in the footsteps of a theological reform, and that theology was nothing else but a university theology"[1] seems to be correct. For example, the first news about the Reformation was usually relayed through humanists and students attached to a university. Of even greater importance than the official attitude of the universities toward the Reformation were friendship cliques that developed among students. As followers of Luther, they recognized each other as partners in a common cause, and were willing to learn from each other. These groups were all the more significant since communication links were still weak, living accommodations scarce, and travelling hazardous. Hence students from a city travelled, matriculated, and often ex-matriculated in groups sometimes consisting of five or six persons. A person in search of higher education usually attended the university closest to his place of residence. Therefore, students from northern Württemberg preferred the university in Heidelberg, students from central Württemberg chose Tübingen, and students from the southern and western tier of Upper Germany attended mostly the universities in Freiburg and Basle. One fact ought to be kept in mind: for the sixteenth century, it is difficult to differentiate sharply between universities and other academic institutions such as Latin schools. A solid education in one of the latter often compared in quality with a short education at a university.

Whereas in Franconia and northern Germany it was the students attending Wittenberg, Leipzig, and Erfurt, who were impressed by Luther's performance in Wittenberg, in Upper Germany proper the students of Heidelberg university

---

[1] Gustav Benrath, "Die Universität in der Reformationszeit," Archiv für Reformationsgeschichte, LVII (1966), 38.

were no less impressed by Luther's debut in Heidelberg on April 26, 1518.[1] From these important cultural and educational central places, Luther's ideas and concepts were diffused by a large number of enthusiasts. The university in Freiburg assumed a similar role, and it was no accident that Heidelberg and Freiburg became diffusion centers for the Reformation before the larger cities of Strasbourg and Constance joined or supplanted them.

Schöffler's assertion that the universities of Mainz, Heidelberg, Freiburg, Basle, and Tübingen vehemently opposed Luther's teachings, and that, therefore, Wittenberg stood completely alone in the fight to carry them through, is not correct.[2] This claim might hold true for Tübingen, and, to some extent for Basle.[3] But to apply it to the Upper German universities is a serious error.[4] It was not merely one university, Wittenberg, that aided the establishment of the new church, for in that case much of the Upper German Reformation could not have acquired its humanistic bent from the universities in Basle, Vienna, Freiburg, and Heidelberg. Schöffler viewed the role of Upper German universities primarily on the basis of the political leadership of the territories in which they were located. Yet it is a fact, that the territorial rulers clamped down on the new ideology only when it began to be popular, that is, from 1522 on. Meanwhile, a good number of people had already accepted it within the framework of their academic institutions. And just as Freiburg and Heidelberg were forced

---

[1] Students also introduced Luther's ideas and writings to foreign universities. In Bologna, for example, German students had a strong impact on their Italian counterparts. See: Bernd Moeller, Johannes Zwick und die Reformation in Konstanz, Quellen und Forschungen zur Reformationsgeschichte, No. 28 (Gütersloh: Mohn, 1961), p. 36. In England, likewise, the university town of Cambridge was the receptacle for Luther's ideas. See: Hans J. Hillerbrand, "The Spread of the Protestant Reformation," p. 276.

[2] Herbert Schöffler, Die Reformation (Frankfurt: Langendreer, 1936), p. 60.

[3] Heinrich Hermelink, Die Theologische Fakultät in Tübingen vor der Reformation (Tübingen: Mohr, 1906), p. 171. Almost all Tübingen humanists and theologians opposed the Reformation. Their removal in 1534 was allegorized with cleaning the "Augean stable."

[4] Accordingly, Schwiebert's contention that there is a striking correlation between the diffusion of the Reformation in different regions and the student enrollment at Wittenberg university, is a strong overstatement. In mapping their place of origin in southwestern Germany, this writer was struck by the dearth of Wittenberg students along the upper Rhine between Worms and Lake Constance on the one hand, and their comparative density on the preacher map. See Ernest Schwiebert, "The Reformation from a New Perspective," Church History, XVII (1948), 20.

to curtail the free exchange of evangelical ideas, Oecolompadius began to teach them at Basle university a year later, following his expulsion from the Ebernburg.

In the Palatinate, with its university at Heidelberg, a free flow of information was tolerated until January 16, 1522, when an important electoral decree was issued against the Wittenberg ideology. The elector had his eyes especially on the students at the university and the preacher's sermons.[1] Similarly, the prince bishop of Mainz ordered the discontinuance of sermons in 1523, and the Austrian government forced the university in Freiburg to take measures against followers of Luther beginning in November 1522.[2] Finally, subjects of territories governed by Austria and Bavaria were prohibited from attending the university in Wittenberg following a decree issued at the Regensburg Convent in 1524.[3]

However, Lutheran sentiments had gained a foothold among a number of students and teachers. Indeed, in many cases, Lutheran friendship cliques had developed prior to Luther's protest in 1517. The most notable one formed in Freiburg, where six to nine reformers were enrolled between January of 1503 and July of 1504, as well as between July of 1514 and October of 1516.[4] These ties continued well into the Reformation period, thereby strengthening the movement and deepening its personal bonds.

Similar patterns of co-operation developed at the University of Heidelberg between Bernhard Wurzelmann, Franz Irenicus, Theobald Billican, Erhard Schnepf, and Jakob Otter during September of 1510 and April of 1512.[5] Several

---

[1] Walter Müller, Die Stellung der Kurpfalz zur lutherischen Bewegung von 1517 bis 1525, Heidelberger Abhandlungen zur mittleren und neueren Geschichte, No. 60 (Heidelberg: Winters, 1934), p. 35.

[2] Winfried Hagenmaier, Das Verhältnis der Universität Freiburg i. Br. zur Reformation (published Ph.D. dissertation, University of Freiburg, 1968), p. 22.

[3] Bruno Gebhardt, ed., Handbuch der deutschen Geschichte (8th ed.; Stuttgart: Union, 1955), II, 50.

[4] They were: Jakob Ofner, Ludwig Oeler, Wilhelm von Fürstenberg, Balthasar Hubmayer, Siegmund Rötlin, Matthias Zell, and Jakob Sturm. See: Hermann Mayer, ed., Matrikel der Universität Freiburg von 1460 bis 1656 (Freiburg: Herdersche, 1907), I, 148-57. The second group consisted of Wolfgang Vogel, Bonifacius Amerbach, Philipp Engelbrecht, Thomas Blarer, Jodocus Hofflich, Jakob Strauss, Sebastian Mayer, Bartholomäus and Konrad Hermann, and Jakob Haystung; ibid., pp. 215-28.

[5] Gustav Toepke, ed., Die Matrikel der Universität Heidelberg von 1386 bis 1662 (Heidelberg: Universitätsverlag, 1884), pp. 477-85.

reformers were still in Heidelberg when Johannes Brenz enrolled in October of 1514. Smaller groups, often consisting of not more than two persons, have been traced elsewhere. However, no such recognizable friendship cliques could be found among Wittenberg students from Upper Germany prior to 1518. This was largely attributable to their limited number due to the long distance between the North and the South. A small number of converted preachers went to Wittenberg, after having been educated in one of the universities in the South and having preached one or two years. These groups can be clearly distinguished on the basis of their predilection either for the Zwinglian-Bucerian or Lutheran teaching. Otter is the only exception within the Heidelberg clique. He also belonged to the Freiburg circle headed by Zasius. Bucer and Beck likewise detached themselves from the Lutheran group, headed by Brenz and Billican.

It is informative to turn away from these individual solicitors of the new faith and focus attention on the attendance patterns of students from 20 cities in the Southwest at the universities in Heidelberg, Wittenberg, and Freiburg. A perusal of Table 3 gives some clues as to which side these cities took in the Luther-Zwingli dispute over the nature of the Lord's Supper. The matriculation dates range from 1515 to 1524, by which time several authorities prohibited their subjects from attending the university in Wittenberg. Those cities with a heavy attendance in Wittenberg and Heidelberg leaned toward the Lutheran doctrine; those scoring light attendance here, but heavy enrollment at Freiburg, chose the Zwinglian branch of the Reformation. Ulm, Memmingen, Lindau, and Giengen appear to be the only exceptions; yet the aspirations of Memmingen and Lindau toward a synthesis of Zwingli's and Luther's beliefs (as exemplified by their membership in the Confessio Tetrapolitana, together with Strasbourg and Augsburg), points to their proximity to Switzerland as well as their students' preference for the universities in Wittenberg, Leipzig, and Heidelberg. Without doubt, location along the key trade route between St. Gallen and Cracow, that went through Leipzig, was an important determinant in this attendance pattern. For example, the Diesbach-Watt (Vadian) trading establishment in St. Gallen had close business ties to cities in Saxony, Silesia, and Poland. In an era when communication was anything but safe and speedy, students often followed in the footsteps of guarded merchants in the selection of their alma mater. Christoph Schappeler, the first Lutheran parson in Memmingen and a native of St. Gallen, who studied and taught at the University of Leipzig between 1499 and 1510, provides a good example of this practice.

In addition, the Peasants' War led to a decline in enrollment at all universities. Its effect was so negative, that in the years immediately following, education at Germany's universities came almost to a standstill. This caused a shortage in the number of available preachers and, ultimately, a slackening of the entire Reformation movement.

Roads and other Communications

Hektor Amman has tried to demonstrate the impact of trade relations and trade routes on the diffusion of the Reformation in western Switzerland. He concludes that merchants were responsible only to a limited degree for transplanting the new ideology.[1] On the whole, this topic has received scant attention in Reformation literature. We know that travelling salesmen often were politically and religiously active, collecting Lutheran literature and news and spreading them in their home-towns and other places where they travelled. Hence, Lutheran gatherings and sermons in towns located along the main roads always drew a better attendance, thanks to the merchants, academicians, master builders, journeymen, beggars, and entertainers.[2]

However, in most cases, a direct connection between trade routes and the dispersion of the new faith is hard to establish. There are almost no data on which highways were chosen by preachers on their itineraries. We are acquainted only with their origins and destinations. Most of the information on trade is restricted to the southern tier of southwestern Germany, an area that received the main impetus for change from Switzerland and the Lake Constance area, which was conspicuous in trading manufactured products over long distances. Merchants in Isny, Wangen, Leutkirch, Biberach, Kempten, Kaufbeuren, Lindau, and Memmingen had intimate business connections with their counterparts in St. Gallen, Zurich, Basle, Constance, and Geneva. The cities of Lindau, Kempten, Wangen, and Isny, moreover, were located along the "Salt Route" that follows the northern rim of the Alps from Hall/Tyrol and Reichenhall in the east, to Zurich in the west. The Zurich-St. Gallen merchants had signed a contract with a number of forwarding agencies in Lindau, the town

---

[1] Hektor Amman, "Oberdeutsche Kaufleute und die Anfänge der Reformation in Genf," Zeitschrift für Württembergische Landesgeschichte, XIII (1954), 150-93; see especially p. 150.

[2] Gustav Bossert, "Kleine Beiträge zur Geschichte der Reformation in Württemberg," Blätter für Württembergische Kirchengeschichte, N. F. VIII (1904), 160.

TABLE 3

STUDENTS FROM VARIOUS CITIES ATTENDING THE UNIVERSITIES OF
HEIDELBERG (H), WITTENBERG (W), FREIBURG (F) 1515-1524

	Augsbg.			Strasbg.			Ulm			Nurembg.			Constance			Esslingen			Hall			Gmünd			Heilbronn			Memming.		
	H.	W.	F.	H.	W.	F.	H.	W.	F.	H.	W.	F.	H.	W.	F.	H.	W.	F.	H.	W.	F.	H.	W.	F.	H.	W.	F.	H.	W.	F.
1515	1		3					3	3		2	1	1		4	1		1	1			4		5	3		1			1
1516	2		3								2	1			5	1		1	2		1	1			5	1	2		2	1
1517	1		3			2	3		1		1			2								1	1	1	1		1	1		
1518	1		5			9	1		5		4	1			2			1	1				1		5					
1519						1					5		1	2	2	1		4		2					3	1			1	2
1520			5	2		8					14	2			6				5					2	5			1	2	
1521	1		10		2	3	1		6		5	1			4				3		1				1			1	1	1
1522	1	1	4		2			2			3			3	5	2	1	2	3				2		5	2				
1523	3				1			1			2						5			1			1						2	
1524					1			1			4																		4	
Total	7	4	33	2	7	22	9	10	9		42	6	2	7	29	5	6	9	15	3	3	6	4	8	28	4	4	4	11	5
	44			31			28			48			38			20			21			18			36			20		

	Nördlingen			Rothenbg.			Stuttgart			Würzbg.			Dinkelsb.			Giengen			Isny			Lindau			Pforzheim			Kempten		
	H.	W.	F.	H.	W.	F.	H.	W.	F.	H.	W.	F.	H.	W.	F.	H.	W.	F.	H.	W.	F.	H.	W.	F.	H.	W.	F.	H.	W.	F.
1515	1	1		1			1						2					1	2		1			1		2				3
1516		1		2				1			2	2					1		2	1		1	2		1	1				4
1517	1					1		1			3	4	1								1	1	2		2			1		
1518								1	1														3	1	1	1				4
1519	2	2									2					1	3			1			1		1	1				4
1520	2	1				1	1				2			1	3				2	2	2			1	1	3				1
1521	1	1				1		1	1	1		1	1	1		1	1		1		2			1	1	2	2		2	
1522	1	1	2	1		1		1	1	3			1							2	2	1	1	1	1					3
1523																				1			2			1				
1524											4																		1	
Total	8	6	6	4		4	1	3	3	4	13	7	5		3	2	5	1	7	3	6	2	11	5	6	7	4	1	4	19
	20			8			7			24			8			8			16			18			17			24		

Source: Carolus E. Foerstemann, ed., Album Academiae Vitebergensis, 1502-1560 (Leipzig: Tauchnitz, 1841); Mayer, ed., Matrikel der Universität Freiburg von 1460 bis 1656; and Toepke, ed., Die Matrikel der Universität Heidelberg von 1386 bis 1662.

which called itself the "protective wall and breadbasket" of the Swiss Confederacy.[1]

A good illustration of this pattern is provided by Joachim Vadian, the foundation preacher and reformer of St. Gallen, who followed closely the Reformation movement in Upper Swabia, across the Lake of Constance. In addition to his personal correspondence, he was kept well-informed by local textile merchants, special observers, and correspondents.[2] The roads from Zurich and St. Gallen to the Upper Swabian cities, among them Ulm and Augsburg, were among the busiest in the entire Southwest. They were surpassed only by roads that paralleled the Rhine Graben further to the west.

The upper Rhine valley between Basle and Frankfurt/Main was, and still is, a thoroughfare of continental importance. The bead-like distribution of preachers, of whom the majority were non-native to their localities, partially verifies the interrelationship between communication and the acceptance of the new theology. For example, it can be shown that the town of Neuenbürg profited from its location along this key route. In July of 1522, its intended preacher, Otto Brunfels, stopped here as he journeyed from Mainz to Switzerland. He made such a positive impression on the leading persons of the town that the magistracy asked him to stay and assume leadership of the local parish.[3]

The most vital thoroughfare in the Black Forest was the Kinzig Valley that separated the sparsely-populated highland into two districts. Its river empties into the Rhine near the metropolitan center of Strasbourg; hence the valley was a convenient and important connection between the central section of the upper Rhine plain, Alsace and France in the west, and the Danube valley, Upper Swabia, and the basin of Lake Constance in the east. Its towns of Offenburg, Gengenbach, and Biberach, as well as Villingen on the eastern side of the mountains, received their main reformatory impulses from Strasbourg. The Aus-

---

[1] G. Meyer v. Knonau, "Zürcherische Beziehungen zur Reichsstadt Lindau," Schriften des Vereins für Geschichte des Bodensees und seiner Umgebung, XLI (1912), 3-13; and G. Meyer v. Knonau, "Die eidgenössische Besatzung in der Reichsstadt Lindau im spanischen Erbfolgekrieg," ibid., p. 40.

[2] Näf, Vadian und seine Stadt St. Gallen, p. 329; and Hans Peyer, Leinwandgewerbe und Fernhandel der Stadt St. Gallen von den Anfängen bis 1520, St. Galler Wirtschaftswissenschaftliche Forschungen, No. 16, Pt. 2 (St. Gallen: Zollikofer, 1960), pp. 32-35.

[3] Karl F. Vierordt, Geschichte der Reformation im Grossherzogtum Baden, 2 vols. (Karlsruhe: Braun, 1847), I, 175-76; and F. Roth, "Otto Brunfels," Zeitschrift für die Geschichte des Oberrheins, IX (1894), 284-320.

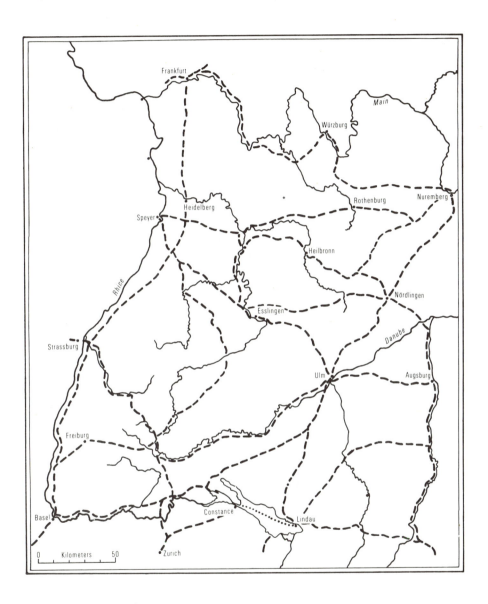

Fig. 14.--Major Roads and Thoroughfares of Southwest Germany

trian government, in a letter to King Ferdinand, complained about the increasing influence in Villingen of the "Lutheran sect originating further west in the Black Forest, as well as other domains (Unterherrschaften) of His Majesty."[1] This communication clearly refers to the Kinzig Valley thoroughfare. The connecting link to the other domains was in all probability the "Swiss Road" between Stuttgart and Schaffhausen.

Another vital road along which the Lutheran message travelled can be traced to the highway from Worms to Nuremberg and Donauwörth.[2] This was one of the three important east-west circuits that saw a large amount of wine from the Rhine and Neckar valleys move to more interior locations, such as Ellwangen, Dinkelsbühl, Feuchtwangen, Ansbach, and Nuremberg. Without doubt, the alliance between Dinkelsbühl and Schwäbisch-Hall, regarding the hiring of Bernhard Wurzelmann from Schwaigern near Heilbronn, was a result of this trade link. Approximately two decades later, Veit Dietrich, a Nuremberg preacher, contacted the preachers Brenz and Gayling in Hall and Weinsberg about the procurement of quality wine from the middle Neckar area.[3]

Mining, Population Mobility, and the Reformation

During the late medieval period, the Black Forest gained in economic importance owing to the presence of rare ores, primarily silver, with copper and lead as valuable by-products. Like other German highlands, this mountainous area began to play a vital role in the development of an early capitalist economy.[4] Moreover, the demand for specialized labor necessary to mine and process these ores led to a busy movement of groups and people between the mountain and hill areas that surrounded the southern plateau.

Because of their scarcity and economic significance, these resources sparked an intense competition among various authorities, and miners enjoyed

---

[1] Ludwig Baumann, "Zur schwäbischen Reformationsgeschichte. Urkunden und Regesten," Freiburger Diözesan-Archiv, X (1876), 108-9.

[2] Karl Weller, "Die Hauptverkehrsstrasse zwischen dem westlichen und südöstlichen Europa," Württembergische Vergangenheit (Stuttgart: Kohlhammer, 1932), pp. 89-129.

[3] Erich Weismann, Zur Geschichte der Stadt Weinsberg (Weinsberg: Haupt, 1959), p. 40.

[4] Erich Otremba, "Wertwandlungen in der deutschen Wirtschaftslandschaft," Die Erde, II (1950/51), 239.

special material and spiritual privileges[1] that demand special attention. Even as late as the Counter-Reformation, the son of Friedrich von Fürstenberg was forced to grant Swiss miners and smelters full religious immunity.[2] Indeed, places as far away as the mining regions of Saxony, Tyrol, the Thuringian Forest, and the upper Palatinate sent some of their surplus laborers into this area.

In approximately 1525, one of the mine owners, Jakob Tenzel from Tyrol, ordered miners from Schwaz, the Tyrolean mining center, and smelterers from Styria into southwestern Germany.[3] The Alpine mining towns of Schwaz, Brixlegg, and Rattenberg, previously had become familiar with Lutheran ideas through contact with miners and merchants from Saxony and Thuringia, where the new faith was tolerated.[4] In fact, the miners in Schwaz, Hall, and Sterzing, by virtue of their professional contacts with their counterparts in central Germany, were the very first Tyroleans to accept Luther's teaching, as early as 1521. In the case of Hall, the miners even physically protected Rhegius on his way to church.[5]

In Galmei near Nussloch/Baden, Jakob Bargsteiner from Amberg in the Upper Palatinate (sometimes labeled as the "Ruhr District" of medieval Germany) acquired special privileges for the establishment of a mining consortium in 1468. Eight years later, he staffed it with miners from the leading mining centers of Goslar and Freiberg. Mines existed also in Bruchsal, Durlach, and Grötzingen. The stock-book in Durlach from the year 1532 mentions the field-name "Silver Mine" in the last two designated localities.[6] Whether or not the mines were worked prior to that date is not known. However, regardless of

---

[1] Max Weber, General Economic History (New York: Collier Books, 1961), p. 143.

[2] The wealth accumulated in the city of Freiburg, displayed in the many attractive houses and churches, is largely a result of the financial gains made by many of its burghers in the nearby silver mines. See Eberhard Gothein, Wirtschaftsgeschichte des Schwarzwaldes und der angrenzenden Landschaften, 2 vols. (Strassburg: Trübner, 1892), I, 664.

[3] Ibid.

[4] Joseph Egger, Geschichte Tirols von den ältesten Zeiten bis in die Neuzeit, 3 vols. (Innsbruck: Wagner, 1880), II, 85.

[5] "Die Glaubenstrennung in Tirol," Historisch-Politische Blätter, VI (1840), 582-83.

[6] "Zur Geschichte des Bergbaues von Nussloch bis Durlach von 1439 bis 1532," Zeitschrift für die Geschichte des Oberrheins, I (1850), 43-48.

their exploitation, workers were certainly engaging a Protestant preacher, or, perhaps, taking advantage of one already residing in these towns. A point in fact is the ore mine at Friesenberg, which was known as the "Evangelical mine" by both its workers and the populace.[1]

The same trend can be observed in the Alsatian mining communities across the Rhine. When Bucer accused the miners of being unable to lead a Christian way of life, the Lutheran preacher in Fortelbach, a mining town near Markirch, expressed his disappointment. He replied: "We also have admirers of the Word of Christ." Most of the miners were immigrants from Saxony, where they had previously embraced the new faith.[2]

When miners began to be recruited from the Upper Palatinate, Duke Friedrich in 1524 allowed this under the condition that among the laborers any agitation against the existing liturgy would be prohibited. Furthermore, strong measures against the new heresy were to be initiated.[3] Whether or not this had the desired effect is questionable, since his successor renounced this ruling for pragmatic reasons.

The fact that miners everywhere joined the new movement with such eagerness and zeal leads one to conclude that they, unlike many other trades, were relatively immune to political authorities. Their special skills played such an important role in bringing prosperity to the cities and towns in south and central Germany that the various governments were forced to tolerate their dissenting religious orientation. The longer the distances covered in their migrations between the various mining regions, the more intensive was their proselytizing of the new doctrine. The fact that Luther himself was the son of a miner probably reinforced this inclination toward the new faith.

---

[1] Katterman, Die Kirchenpolitik Markgraf Philipp I, p. 42.

[2] Adam, Evangelische Kirchengeschichte der Elsässischen Territorien, p. 350.

[3] Gothein, Wirtschaftsgeschichte des Schwarzwaldes, I, 664.

CHAPTER IV

POLITICAL DECISION-MAKING

AND THE REFORMATION

Hanns Rückert has pointed out that ecclesiastical movements lacking any connection to political organization are usually not among the strongest and most significant events leading to religious change.[1] This applies particularly to sixteenth century Germany, in which the primitive identity of religious and political affiliation was still unbroken. In spite of the breakup of a unified belief system by the Reformation, this twin identity of church and state persisted for a long time. It meant that the secular ruler had authority over the faith of his subjects; not that he had the power to create or determine faith, but rather that he had the right and duty to prohibit and punish a public faith which he considered wrong.[2] This is precisely the meaning of cuius regio eius religio, i.e. the religion of the ruler shall be the religion of his territory.

In the fifteenth and sixteenth centuries, the ecclesiastical power of bishops experienced a rapid decline in the face of the increasing significance of territorial rulers and their cities. Hence, political figures such as Emperor Charles V and Archduke Ferdinand of Württemberg were the strongest political supporters of the bishop of Constance. Whereas the majority of territorial rulers clamped down on the Reformation in order to avoid upheavals, which might result in the loss of their newly acquired political control, the majority of magistracies were cautious not to impede the movement for fear of serious civil disturbances. The inability of the various administrations to arrive at a common definition of law and order, ultimately delayed the Reformation by more than one decade.

---

[1] Hanns Rückert, "Die Bedeutung der württembergischen Reformation für den Gang der deutschen Reformationsgeschichte," Blätter für Württembergische Kirchengeschichte, N.F. XXXVIII (1934), 268.

[2] Johannes Heckel, "Cura religionis, ius in sacra, ius circa sacra," in Festschrift Ulrich Stutz, Kirchenrechtliche Abhandlungen, No. 17-18 (Stuttgart: Enke, 1938), p. 234; this argument was used by weakened Catholics or Protestants as soon as their opponent was strong enough to expand spatially at the expense of the other.

The traditional enmity between cities and principalities was not helped by the progress and success of the movement in urban areas. In political terms, the secular princes were the most powerful administrators of the German estates. The relationship between church and city likewise was one of enmity, but here the cities were often able to score a victory over ecclesiastical authorities, provided that the territorial rulers did not side with the latter. To be sure, a city's opposition to church authority within its walls sometimes was as important a factor for religious defection as a genuine belief in the new faith. For example, Strasbourg, Speyer, and Landau were cities that experienced very strained relations with their bishops.

Unlike the urban areas, the villages or parishes in small knightly territories depended on political leadership for their choice of faith. The knights either hired preachers out of the universities or from other territorial cities and towns.

The highly fragmented territorial situation in sixteenth century Germany poses difficulties for a comprehensive understanding of how the Reformation proceeded in a spatial context. Hence, most research has been carried out on a local or territorial basis; it treats these small territories as autonomous systems, thereby failing to see the forest for the trees. This orientation toward political, theological, and canonical problems has encouraged a particularistic approach, one that omits connective links and external influences.[1] The Southwest, not unlike the area along the lower Rhine and Polish-German border was often characterized by a confusing pattern of tiny, dispersed territories. The political map shows a division into innumerable juxtaposed and even overlapping city states and towns, knightly and princely territories, as well as ecclesiastical provinces that included the larger prince-episcopal components (Fig. 15).

Imperial Cities

Imperial cities and towns, in contrast to territorial ones, occupied a special position, since they existed as parts of the Holy Roman Empire rather than of emerging political territories. Unlike knightly territories, they acquired the

---

[1] This approach has also been applied to larger territories in Germany. See Kurt Dietrich Schmidt, "Die konfessionelle Gestaltung Deutschlands. Nichttheologische Faktoren bei Separationen und kirchlichen Zusammenschlüssen," Theologische Literaturzeitung, LXXVII (1952), 134; and Wilhelm Sievers, Über die Abhängigkeit der jetzigen Confessionsverteilung in Südwestdeutschland von den früheren Territorialgrenzen (Göttingen: Peppmüller, 1884).

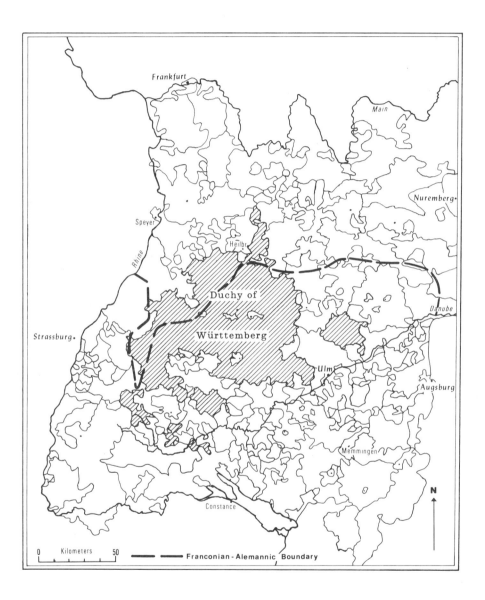

Fig. 15.--The Political Division of Southwest Germany at about 1500

privilege to participate in the Imperial Diet. And unlike sovereign and aristocratic territories based on authoritarian principles, the imperial cities also played a mediating role between authoritarian and co-operative political systems.[1]

On the political map they appear like cavities in the surrounding territories, although several of them such as Ulm, Rottweil, Schwäbisch-Hall, Rothenburg/Tauber, possessed such large chunks of land that they equalled some of the medium-sized territories. Nominally they were subject to the Emperor. Yet, in light of the centrifugal tendencies that gained dominance within the Empire, they constituted independent political entities.[2]

Of the 34 imperial cities in the German Southwest only three, Zell am Hermersbach, Buchau, and Pfullendorf, were completely unaffected by the Reformation. This fact proves that the Reformation had a great impact on cities that were comparatively free from political pressure. Furthermore, the prominent role which some of them played in the diffusion of the new religion can not be overestimated, as Strasbourg, Constance, Memmingen, Ulm, and Schwäbisch-Hall clearly illustrate.

Knightly Territories

Knights contended with imperial cities as followers and distributors of Luther's ideas. The small knightly territories often appeared in conglomerates that formed a buffer zone between larger territories, such as the Knights Canton of Kraichgau, located between the Duchy of Württemberg, the Margraviate Baden, and the Electorate of Palatinate. All three of them tried to dominate this area. Unlike the larger territorial states, the knightly possessions lacked larger urban settlements, and were therefore almost completely rural in character. Because of their limited size, they often permitted close contact be-

---

[1] Karl S. Bader, Der deutsche Südwesten in seiner territorialgeschichtlichen Entwicklung, p. 149.

[2] Bernd Moeller, Reichsstadt und Reformation, Schriften des Vereins für Reformationsgeschichte, No. 180 (Gütersloh: Mohn, 1962); including his shorter article, "Die Kirche in den evangelischen freien Städten Oberdeutschlands im Zeitalter der Reformation," Zeitschrift für die Geschichte des Oberrheins, CXII (1964), 147-62. The first work has been partly translated in Bernd Moeller, Imperial Cities and the Reformation: Three Essays, trans. and ed. by H. C. Erik Midelfort and Mark U. Edwards (Philadelphia: Fortress Press, 1972), pp. 41-115.

tween ruler and ruled.[1]

The Franconian knights in particular showed great interest in Luther. For example, Ulrich of Hutten enthusiastically adopted and spread the slogan "Away from Rome!" He became interested in Luther because the Wittenberg reformer offered a German aura that reflected the accumulated anti-Roman feelings of a large percentage of German knights. For them Luther symbolized "The German" simply because he was anti-Roman.[2]

Another Franconian knight, Franz of Sickingen, offered Luther and other reformers protection on his Ebernburg.[3] At first, Sickingen was not impressed with Luther's stand at Worms, and it was up to Hutten to convert this fellow knight to Lutheranism. We have here a typical case of a disciple going to the source (Leipzig Debate, 1518, and Wittenberg), and spreading the ideology with his return. Hutten can be credited with introducing Sickingen to the secondary reformers as well.[4]

Two knights, Hans Landschad of Neckarsteinach and Dietrich of Gemmingen, became converts as a result of Luther's presence in nearby Heidelberg and in Worms. Landschad, whose two sons studied in Heidelberg, had close ties to Martin Bucer there and later in Strasbourg. This knight had received the new faith in 1520 or earlier, before he sent a letter to the Saxon elector expressing his support for Luther.[5] He was a Burgmann (or high city official) in Oppenheim where Luther stayed for two nights on his way to and from Worms in April, 1521. Here Luther met Bucer at the court of Hartmut of Kronberg in the presence of Landschad. It was Bucer who procured a preacher for Landschad's town of Neckarsteinach in 1525.[6]

---

[1] Karl S. Bader, "Reichsadel und Reichsstädte in Schwaben am Ende des alten Reiches," in Aus Verfassung und Landesgeschichte, Festschrift Theodor Mayer (Konstanz: Thorbecke, 1958), pp. 259-60.

[2] Lortz, The Reformation in Germany, I, 248.

[3] Erwin Riedenauer, "Reichsritterschaft und Konfession," in Deutscher Adel 1555-1650, ed. by Helmut Rösler (Darmstadt: Wissenschaftliche Buchgesellschaft, 1965), pp. 5-6.

[4] Hitchcock, The Background of the Knights' Revolt 1522-1523, p. 10.

[5] B. Berbig, "Ein Brief des Ritters Hans Lantschad zu Steinach an Kurfürst Friedrich den Weisen, 1520," Archiv für Reformationsgeschichte, II (1905), 391.

[6] Steitz, Geschichte der evangelischen Kirche in Hessen und Nassau, I, 37.

Dietrich of Gemmingen became the opinion leader of the Kraichgau knights after his return from the Diet at Worms.[1] He was well educated and appointed a number of Heidelberg university graduates as evangelical preachers to village churches of which he had the patronage. For example, he asked Bernhard Griebler to preach in Gemmingen and Berwangen, Erhard Schnepf at Guttenberg Castle, Heinrich N. in Bonfeld, and Martin Germanus in Fürfeld. Since the Gemmingen preacher was at the same time superintendent of the Latin School, he cast a strong influence over the neighboring knights whose sons received their education there. Dietrich and his preachers always maintained close ties to Brenz in Schwäbisch-Hall, who was the central figure of the intra-Protestant communion dispute between the Rhine and Nuremberg.

The preference which many burghers of the town of Wimpfen showed toward Luther's teaching must be attributed primarily to the influence of neighboring knights, including those in cities with whom the imperial city had close ties. It was chiefly the Gemmingen knight, together with the Ravensberg, Degenfeld, Berlichingen, Göhler, and Sickingen families, who was responsible for the promotion of the new faith in Wimpfen. Their territories adjoined that city to the west and north; and since they constituted the hinterland with which the city exchanged goods, Wimpfen functioned simultaneously as capital of the Knight's Canton. The lower nobility had property and enjoyed burgher status in the imperial town. When Gemmingen hired Schnepf as preacher for Guttenberg Castle and Neckarmühlbach, the latter also delivered sermons in Wimpfen.[2]

Most of the neighboring knights followed Gemmingen's example. In 1522, von Ravensberg hired Johann Gallus as parson of Sulzfeld. Like Gemmingen's preachers, Gallus came from Heidelberg where he had enjoyed cordial ties with Brenz.[3] In Fiehingen, Nicolas Trabant, in Neckarbischofsheim, Nikolaus Renneisen and Melchior Hippovius, and in Flinsbach, Nikolaus Thoma were appointed by their knightly ruler. At least two of these preachers attended Heidel-

---

[1] Beschreibung des Oberamts Heilbronn, 2 vols. (Stuttgart: Kohlhammer, 1903), II, 272.

[2] Karl Heid, Die Geschichte der Stadt Wimpfen (Darmstadt: Müller, 1936), p. 116.

[3] Gustav Bossert, "Beitrag zur badisch-pfälzischen Reformationsgeschichte," Zeitschrift für die Geschichte des Oberrheins, N. F. XVII (1902), 81; and Gustav Bossert, "Die Syngrammatisten," Blätter für Württembergische Reformationsgeschichte, VII (1892), 20.

berg university, while a third, Trabant, was a close friend of Brenz.[1]

Schwaigern, which was in the possession of the Neipperg clan, hired, between 1525 and 1528, Bernhard Wurtzelmann from Wimpfen. He later left for Dinkelsbühl. And in Bönningheim, the rulers ordered monks to "preach the gospel and word of God in a pure and clear manner."[2] The famous knight and peasant leader Götz of Berlichingen requested a preacher for Neckarzimmern sometime before 1524, most likely in 1522. While the knight was influenced by his friend Sickingen, his preacher Georg Amerbacher personally communicated with Brenz.[3]

Knight Landschad in Neckarsteinach[4] appointed Jakob Otter as preacher, after the latter was recommended to him by the city of Strasbourg. Hired in 1525, he did not become the town preacher until 1527.[5] The knights Georg and Engelhard IV at Hirschhorn, who ruled five miles upstream on the Neckar, soon followed Landschad's example. They appointed two evangelical preachers, Josias Forster and Jost Butt, in 1528, but they were removed after Emperor Charles V applied political pressure on the knights. When we look at the order of appearance of the preachers in Heidelberg, Neckarsteinach, and Hirschhorn (Figs. 5 and 7), an interesting pattern emerges. What we have here is a rather lucid example of expansion-diffusion, beginning in Heidelberg and gradually moving upstream along the Neckar River. We know that this rather narrow, steep-sloped valley route was heavily used by watercraft as well as land vehicles. Most of them transported fruit and wine from the open Rhine valley to more interior sections of Franconia.[6]

---

[1] Vierordt, Geschichte der Reformation im Grossherzogtum Baden, I, 237; and Toepke, ed., Die Matrikel der Universität Heidelberg von 1386 bis 1662, I, 451, 487.

[2] Gustav Hoffmann, "Reformation im Bezirk Besigheim," Blätter für Württembergische Kirchengeschichte, XXXVIII (1934), 151; and Wolfgang Schmid, Bönningheim: Stadt und Ganerbiat (Tübingen: Institut für geschichtliche Landeskunde, 1969), p. 27.

[3] D. Heinrich Neu, Pfarrerbuch der evangelischen Kirche Badens von der Reformation bis zur Gegenwart, Veröffentlichungen des Vereins für Kirchengeschichte in der evangelischen Landeskirche Badens (Lahr: Schauenburg, 1938), I, 207.

[4] He was the cousin of Dietrich of Gemmingen; the latter established Landschad's excellent rapport with Brenz.

[5] Steitz, Geschichte der evangelischen Kirche in Hessen und Nassau, I, 37.

[6] Goods between Heidelberg and Mosbach were duty-exempt.

Of all knightly territories, the one with the most influence was that of Franz of Sickingen in the Palatinate. His Ebernburg Castle was a center that not only attracted and diffused the new ideology, but was also innovative in the liturgical field. The reason why Sickingen could play such an effective role lay in his personal outlook as well as his economic circumstances. He was pious; yet he was able to increase his wealth through the mining of scarce ores in the Electorate of Palatinate. This allowed him plenty of leisure time to think about economic and political issues, and, ultimately, to support various persons who were to become great names in the modern religious movement.[1]

Two of the more influential preachers in the German Southwest, Martin Bucer and Johann Oecolompadius, early sought refuge at the Ebernburg. Eventually, both were to play extremely active roles in the Reformation movement, first in Strasbourg and Basle, and later in a number of upper German cities. Other preachers who found shelter there included the less well-known Caspar Aquila, Johann Schwebel, Johann Metzler, Nikolaus Merxheimer, Eberlin von Günzburg, and Hartmut von Kronberg.

After Bucer's meeting with Luther in Heidelberg, he suffered a period of increasing alienation from his Dominican order; he left Heidelberg and, after a short service for the Palatinate ruler, became the first Protestant preacher for Sickingen at Landstuhl in May, 1522.[2] Like other preachers, he married, a sure sign that he had severed all previous ties to the old church. Following the defeat of Sickingen in 1523, Bucer escaped to neighboring Weissenburg, then went to Strasbourg. Oecolompadius immediately began to read the daily mass, epistles, and evangelism in German when he arrived in the summer of 1522. It was the first time that this had been done in southwestern Germany, and was considered such a sensation that Kaspar Hedio in Mainz immediately wanted to know more about it. Oecolompadius' next move was to administer communion under both forms.[3] No other locality or city in the Southwest, except Augsburg exhibited such boldness at such an early date.

In 1521, Johann Schwebel arrived from Pforzheim, and Kaspar Aquila from Augsburg. Schwebel became the lifelong superintendent of the duke in Zweibrücken, while Aquila fled to Thuringia after Sickingen's defeat. All of Sickingen's preachers actively participated in numerous conferences with leading Protestant knights and preachers who made visits to the Ebernburg.

---

[1] Heinrich Steitz, "Franz von Sickingen und die reformatorische Bewegung," Blätter für die Pfälzische Kirchengeschichte, XXXVI (1969), 22.

[2] Ibid., p. 25.

[3] Ibid., p. 26.

If Sickingen's influence on Strasbourg, Basle, Zweibrücken, Thuringia, and other places can be said to be only indirect, the knight had a more direct impact on four communities east of the Rhine: Sickingen, Neuenbürg, Rohrbach (near Eppingen), and Wildbad. These he provided with Protestant preachers soon after the religious change had occurred on the Ebernburg.

In the Ortenau area across the Rhine from Strasbourg, a few knights were sympathetic to the Protestant cause. They enjoyed very close ties to the large metropolis. Egenolf von Röder, who acted as Stättmeister (noble member of the governing body) in Strasbourg, enthusiastically supported the Reformation. His villages of Oberschopfheim and Diersburg witnessed the first evangelical sermons around 1530, while Hofweier followed in 1534.[1] Ludwig of Böcklinsau, also a burgher of Strasbourg, attempted to introduce a preacher of the new faith. However, neighboring cloisters, whose territories surrounded his villages, made this impossible. It was always difficult to introduce the new faith in such areas.

The lower nobility that was scattered in and around the Duchy of Württemberg joined the Reformation movement, but not in such a collective spirit as in the Kraichgau and Ortenau. Here, too, affiliations with a city or town with a vigorous preacher seemed almost a prerequisite. Most knightly territories on the upper Neckar, the upper Danube, and along the eastern edge of Württemberg lacked such connections to an important Reformation center. At the same time, they were surrounded by staunchly Catholic areas that hindered religious change.

Among the scattered knightly territories inside Württemberg, Hans Friedrich Thumb of Neuburg took the lead in Köngen, near Esslingen. In 1527, his parson was asked to appear before the episcopal official, who had accused him of "joining the Lutheran sect."[2] Hans Conrad Thumb, in Stetten north of Esslingen, followed in his brother's footsteps. Both had personal ties to the imperial city of Esslingen where they owned a house. In 1532, the Köngen knight, with the support of Ambrosius Blarer, abandoned the mass altogether and removed the images from his church. Blarer was staying in Esslingen at the time, where he assisted the new congregation in the "abolition of the last Roman-Cath-

---

[1] Manfred Krebs, "Politische und kirchliche Geschichte der Ortenau," Die Ortenau, XVI (1929), 142-43; and Neu, Pfarrerbuch der evangelischen Kirche Badens von der Reformation bis zur Gegenwart, I, 152.

[2] Württembergische Kirchengeschichte (Calw: Calwer Verlagsverein, 1893), p. 322.

olic vestiges."[1]

Knight Philip of Rechberg granted Martin Cless asylum on Ramsberg Castle in 1529. He was persecuted for preaching heretical reformatory ideas in Oberhofen outside of Göppingen.[2] Yet, his exile was short. He soon left for Kassel to serve as secretary to Duke Ulrich, who recommended him to Zwingli and Oecolompadius.[3]

Less is known about an active movement in the knightly territories on the upper Rems-Kocher valleys and the upper Neckar. One reason for this is that the movement in the adjacent imperial cities of Schwäbisch-Gmünd and Aalen was squelched too soon to have an immediate effect on the surrounding vicinity. The situation in the knightly territories on the upper Neckar between Horb and Tübingen was in many ways similar. They, too, were hemmed in by larger Roman-Catholic polities.

Princely Territories

During the first half of the sixteenth century, the political and cultural process of extending and solidifying the princely territories was still in full force. The rulers attempted to expand their political power to all aspects of life, including religion. No fact illustrates the impact of a ruler on the cultural morphology of his realm better than the principle of cuius regio, eius religio, i.e., the religion of the ruler determines the religion of the subjects.[4] Both Roman-Catholic and Protestant rulers tightened their grip on their possessions by imposing their line of thought on their populations. In the first decade of the Reformation, however, a clear division into evangelical and Catholic can not be applied to the ecclesiastical policy of German rulers.[5] Generally, their support for the Reformation was weak, especially among the counts, and their attitude toward Lutheranism was characterized by cautious reservation or re-

---

[1] Otto Schuster, Kirchengeschichte von Stadt und Bezirk Esslingen (Stuttgart: Calwer, 1946), pp. 176-77.

[2] David F. Cless, Versuch einer kirchlichen politischen Landes- und Culturgeschichte von Würtemberg bis zur Reformation, II (Schwäbisch-Gmünd, 1808), 185.

[3] Theodor Schön, "Meister Martin von Uhingen als Prediger in Rottenburg a.N.," Reutlinger Geschichtsblätter, XIII (1902), 30.

[4] Hermann Aubin, Theodor Frings, and Josef Müller, Kulturströmungen und Kulturprovinzen in den Rheinlanden (Bonn: L. Röhrscheid, 1926), pp. 53 ff.

[5] Kattermann, Die Kirchenpolitik Markgraf Phillipp I, p. 42.

jection.[1]

As the princely territories varied greatly in size, population, and economic structure, so they also varied in influence. Nevertheless, one must be careful not to equate smallness with restricted influence.[2] A far more important criterion was the existence of a nearby diffusion center with powerful personalities whose charisma radiated into the neighboring territories and beyond. A territory that lacked such a center, remained dependent on "foreign" centers, regardless of its size. In this sense we have "giving" and "receiving" territories.

This rather fixed situation was counteracted by the publication of volatile political decrees resulting in an exodus of preachers that was often just short of a mass emigration. Like a great many knightly territories, princely territories received the impulse to reform from within themselves or from nearby cities where the movement was already well developed.

The Reformation in the Principality of Brandenburg-Ansbach relied heavily on the advice of such cities as Nuremberg, Schwäbisch-Hall, and Crailsheim. Margrave Georg and his brother Casimir ruled the principality; Casimir was staunchly opposed to Lutheranism, while Georg had favored it since 1520.[3] Things changed in 1524-25, when Casimir gave the green light to the worship of God's Word. But any further changes were delayed under strong pressure from conservative estates until Casimir's death in 1528. After that, all the political power was transferred to Georg, who paved the way for the successful introduction of the communion in both forms and abolition of the mass.[4]

The Principality of Öttingen south of Ansbach was ruled by a sympathizer of the new faith, Count Karl Wolfgang, who appointed Paul Warbeck as his preacher. Like Casimir of Ansbach, the count refused to abolish the old order completely. However, he permitted the worship and proclamation of God's

---

[1] G. Haselier, "Die Auswirkungen der Territorialisierung auf die politisch-kulturelle Struktur in der Pfalz und im unteren Elsass," in Probleme der Geschichte und der Landeskunde am linken Oberrhein, ed. Franz Arthen (1966), p. 93.

[2] Friedemann Merkel, Geschichte des Evangelischen Bekenntnisses in Baden von der Reformation bis zur Union, Veröffentlichung des Vereins für Kirchengeschichte in der evangelischen Landeskirche Badens, No. 20 (1960), p. 11.

[3] Franz Herrmann, Markgrafen-Büchlein (Bayreuth: Mühl, 1902), p. 52.

[4] Ibid., pp. 55-56.

Word. Eventually, this was followed by communion in both forms and the substitution of German for Latin in the liturgy.[1] Ansbach and Öttingen were the only territories east of the Duchy of Württemberg whose rulers embraced the Lutheran doctrine prior to 1534.

In the County of Wertheim, the Reformation was popular partly for political reasons. Earl Georg, situated between the two large prince-bishoprics of Mainz and Würzburg, tried to diminish the influence of these powers by encroaching on their ecclesiastical jurisdiction whenever he could. At the Diet in Worms, the earl asked Luther to send him a preacher from Wittenberg. Thus, with the gradual initiation of the Reformation, he made his territory into a free estate responsible only to the Emperor.[2]

In the Palatinate, Elector Ludwig V was considered one of Luther's sympathizers at the Diet in Worms. But this may have been a response to the hatred he felt for Emperor Charles V, rather than a genuine appreciation of Luther's religious stand.[3] He returned to a pro-Catholic position in 1526, out of fear that the decline of the church and the emperor might cause his own downfall. Until then, he had allowed only weak reformers in his territory, while the more outspoken ones (Brenz and Billican) were expelled.[4]

Further south, on the right side of the Rhine, the Margraviate of Baden was split into two parts in 1515. The northern section went to Philipp, and the southern part to Ernst. The latter showed only moderate support for the new faith, while Philipp gravitated toward it between 1524 and 1531,[5] but retracted afterward. None of these brothers signed the Augsburg Confession, partly because they feared political reprisals from the neighboring Hapsburg empire.[6]

---

[1] Friedrich Zoepfl, Das Bistum Augsburg und seine Bischöfe im Reformationsjahrhundert (Augsburg: Winifried-Werk, 1969), II, 52.

[2] Emil Ballweg, "Einführung und Verlauf der Reformation im Badischen Frankenland" (unpublished Ph.D. dissertation, University of Freiburg/Br., 1944), pp. 51-60.

[3] Paul Kalkoff, "Die Prädikanten Rot-Locher, Eberlin und Kettenbach," Archiv für Reformationsgeschichte, XXV (1928), 140.

[4] Ballweg, "Einführung und Verlauf der Reformation," p. 41.

[5] Neu, Pfarrerbuch der evangelischen Kirche Badens von der Reformation bis zur Gegenwart, I, 324-25.

[6] In the words of one author: "Far from wholeheartedly embracing the new faith, neither was it in line with Roman-Catholicism." See Kattermann, Die Kirchenpolitik Markgraf Philipps I, p. 42.

Count Philipp III of Hanau-Lichtenberg, whose territory stretched along the Rhine opposite Strasbourg, became an adherent of the Reformation in 1525. One year later he ordered the abolition of the mass for the dead, and asked the Strasbourg magistracy to dispatch its provost and former preacher Wolfgang Capito to his territory. However, developing political differences between the count and the city of Strasbourg caused him to reject the evangelical creed.[1]

Count Wilhelm of Fürstenberg encouraged early reformatory attempts among his constituents in the Kinzing valley and in the Ortenau.[2]

The rulers of the remaining southwest German principalities opposed religious change in varying degrees. For example, the Duchy of Württemberg and the Hapsburg dominions that were scattered across the southern portion of Upper Germany were forced to remain strictly Catholic. The Principality of Fürstenberg, governed by Friedrich von Fürstenberg, and the counties of Waldburg, Limpurg, and Hohenlohe were some of the larger territories whose ruler passionately eschewed religious change.

Ecclesiastical Territories

The bishoprics in Strasbourg, Constance, Augsburg, Speyer, Worms, Würzburg, and Eichstätt, as well as the archbishopric in Mainz, were in varying degrees and times opposed to the Reformation movement. Here, the bishops were the sovereign authorities, and their territories were more likely to cling to the old faith than the non-ecclesiastical ones. Unlike the bishopric, the diocese covered a territory that was much larger and was made up of various polities such as imperial cities, duchies, and imperial knightly territories discussed in the preceding pages. What concerns us here are the so-called prince-bishoprics.

In 1521, the bishop of Strasbourg requested instant repressive measures against Matthäus Zell, who was the local foundation preacher whose sermons were regarded as undisciplined and inflammatory. Nonetheless, since the preacher was appointed by the cathedral chapter, many of whom rejected their superior's charge, Zell was able to stay.[3]

The Bishop of Constance was somewhat more susceptible to Luther's

---

[1] Krebs, "Politische und kirchliche Geschichte der Ortenau," p. 140.

[2] Ibid., p. 143.

[3] Miriam U. Chrisman, Strasbourg and the Reform: A Study in the Process of Change (New Haven: Yale University Press, 1967), pp. 100-101.

teaching. He expressed hope that the Lutheran movement would infuse new spirit and energy into a deteriorated church. As an opponent of the indulgence practices, the bishop even enlisted Zwingli's support. Not until 1522 did he take action against the preachers in the city.[1]

The Bishop of Augsburg, Stadion, likewise aimed at a church free of corruption and mismanagement, but his attempts to create a better atmosphere were foiled by his close relationship with the banking house of Fugger, which was deeply involved in the sale of indulgences. Beginning in October 1520, the bishop assumed an active role in suppressing the Lutheran movement.[2]

Like Stadion in Augsburg, Bishop Georg of Speyer rejected Luther's ideas right from the beginning, as was shown earlier. As the number of pro-Lutheran pamphlets increased, in the wake of Pope Leo X's Bull Exsurge domine, the bishop was forced to take stringent measures against the clergy who read them to their congregations.[3] In an open letter published and circulated in November 12, 1521, the bishop, for the first time, sharply criticized the disobedience of the "erroribus Lutheranis."[4] Still, despite the reprimand from their bishop, many of the chapter clergy continued to bring Luther's writings to their church.[5]

Even under the aegis of the succeeding bishop, Philip II (1529-1551), many villagers exhibited a strong liking for the new teaching. They visited the reformed worship in villages and towns to the west and south of the bishopric.[6]

Archbishop Albrecht of Mainz, though no ardent supporter or spiritual

---

[1] Hans C. Rublack, Die Einführung der Reformation in Konstanz von den Anfängen bis zum Abschluss 1531, Veröffentlichungen des Vereins für Kirchengeschichte in der evangelischen Landeskirche in Baden, No. 27 (Karlsruhe, 1971), p. 16.

[2] Zoepfl, Das Bistum Augsburg und seine Bischöfe im Reformationsjahrhundert, p. 15.

[3] Karl F. Lederle, "Die kirchlichen Bewegungen in der Markgrafschaft Baden-Baden," Freiburger Diözesan-Archiv, XVIII (1917), 383.

[4] Manfred Krebs, Die Protokolle des Speyerer Domkapitels, Veröffentlichung der Kommission für geschichtliche Landeskunde in Baden-Württemberg, Reihe B, No. 21 (Stuttgart: Kohlhammer, 1966), p. 70, footn. 40.

[5] Franz X. Remling, Das Reformationswerk in der Pfalz (Mannheim: Götzische, 1846), p. 58.

[6] Günter Haselier, Geschichte des Dorfes und der Gemeinde Weiher am Bruhrain (Weiher: Gemeindeverlag, 1962), p. 77. Inhabitants of Weiher, Ubstadt, Kislau, Stettfeld, and Langenbrücken attended services in knightly territories further west.

leader of the old church, continued to be loyal to the Catholic faith. However, as the popularity of Luther grew at his court and among the university faculty, so did the possibility of the bishop's disobedience to the old order.[1] The installation of Wolfgang Capito in 1520 as cathedral preacher, and his appointment to the council was of great significance to the reform movement in Mainz and the entire diocese. Capito, influential with leading personalities at the court, was nevertheless unable to convert the highest authority. Even so, the archbishop tolerated the Lutheran movement for almost three years. Following the decision of the diet on March 6, 1523, he issued a decree six months later, in which he prohibited Lutheran sermons and literature in the diocese. Gradually, the Reformation there came to a halt,[2] although a brief revival occurred during the peasant rebellion. Its suppression was accompanied by another order of the bishop, banishing Lutheran preachers from the entire diocese.[3]

The newly elected bishop of the Würzburg diocese immediately eastward of Mainz, tried to squash the Reformation movement in its roots. But, like the Bishop of Strasbourg, he was hampered by dissent within the cathedral chapter. The defeat of the peasants in 1525 provided him the chance to thwart the movement. He ordered many evangelical preachers or "Lutheran rascals" around Würzburg to be arrested, forcing them to retract.[4]

The bishop of Worms was unable to stop the movement. The imperial city had been witness to long-time differences between the magistracy and the bishop, and finally voted for secession from the old church.[5] The bishopric of Worms was the smallest in southwestern Germany.

Finally, the Eichstätt bishopric, which was almost completely surrounded by the territory of Ansbach, successfully arrested the Reformation in that town. The rest of the ecclesiastical institutions consisted of imperial cloisters (Reichsabteien), the Teutonic order, and the Order of St. John's, all opposed to

---

[1] Andreas Veit, Kirche und Kirchenreform in der Erzdiözese Mainz im Mittelalter der Glaubensspaltung und der beginnenden tridentischen Reformation (1517-1518), Erläuterungen und Ergänzungen zu Janssens Geschichte des deutschen Volkes, No. 10, Pt. 3 (Freiburg: Herder, 1920), p. 18.

[2] Steitz, Geschichte der Evangelischen Kirche in Hessen und Nassau, I, 3-4.

[3] Veit, Kirche und Kirchenreform in der Erzdiözese Mainz, p. 12.

[4] Matthias Simon, Evangelische Kirchengeschichte Bayerns (Munich: Müller, 1942), p. 212.

[5] Remling, Das Reformationswerk in der Pfalz, p. 58.

the Reformation.

The most threatening enemy of the Protestant-oriented imperial cities was the Swabian Federation. It actively supported the bishop of Constance in subduing the reform movement in his area.[1] The stronghold of the Federation was in the rural sectors of the East and the South. It was weakest toward the Palatinate and the upper Rhine, and generally, wherever the Palatinate made its power felt. This explains the dearth of evangelical preachers in the belt of rural territories that formed an arc stretching from the southern Black Forest northeastward to the imperial city of Rothenburg. It may also account for the sudden halt of the Reformation movement in the imperial cities within this region. On the other hand, the movement so weakened the Federation in the middle twenties that it found itself virtually powerless after 1528-29.[2]

In summary, the wave of political reformations that began in the 1530's was accompanied by important political decision-making during that period. The rejection of Lutheranism by rulers was chiefly the result of political considerations rather than private soul-searching. Still, it led to a constant shifting of preachers from one territory into another, and often back again. For example, prior to 1532, many preachers in Württemberg sought refuge in Baden, a trend which was reversed a short time later.

The territorial reformation in southwestern Germany was limited to the northern or Franconian section. This area was closest to and shared some of the characteristics of northeastern Germany. Moreover, it depended on Wittenberg, Leipzig, and Erfurt for its supply of preachers. The territory of Ansbach (Öttingen and Wertheim to a lesser extent) was like the northeastern German states in that it was larger in size and less developed in its economy. The Palatinate and the knightly territories of the Kraichgau were "infected" mainly through the Rhine corridor and its capital city Heidelberg. In contrast, the Reformation movement in the southern, or Alemannic sector of southwestern Germany was limited primarily to cities and towns.

Special emphasis has been placed on the knightly territories in the upper Rhine area because of their crucial role as initial sympathizers and disciples of the new religion. The impact of some of its noblemen was so strong that they

---

[1] August Willburger, Die Konstanzer Bischöfe Hugo von Landenberg, Balthasar Merklin, Johann von Lupfen (1496-1537) und die Glaubensspaltung, Reformationsgeschichtliche Studien und Texte 34-35 (1917), pp. 213 ff.

[2] According to Prof. Decker-Hauff, Tübingen.

determined the course of the Reformation in a number of cities, including Basle, Wimpfen, Zweibrücken, Strasbourg, and Dinkelsbühl. Indeed, the residences of Franz of Sickingen and Gemmingen served more prominently as innovation and diffusion centers than many cities.

CHAPTER V

PLACES OF ORIGIN OF PREACHERS AND LOCALITIES
WITH RESIDENT PREACHERS

The preachers commissioned to a certain locality between 1518 and 1534 can be divided into two groups: (1) resident preachers[1] and (2) non-resident preachers. In the first category belong those clergy who preached the Gospel in the same locality where they were born. In the second category are those who were born in another place than where they preached. The larger, non-resident group can further be subdivided into (a) persons who began to preach the new ideas immediately after their arrival in a non-native community, and (b) persons who began to preach the new ideas some time after their arrival in a new community. While some had moved to new locations before 1517-18, and some after that, both groups made their move independent of the religious cleavage.

A map showing place of birth and preaching activity is a convenient way to present a general picture of the communication network of the time. Connecting the various informational data by means of arrows, the author was able to detect several important "bundles" of movements, such as the one along the upper Rhine, connecting Basle with Strasbourg and the Rhine-Main district, between Zurich-Constance and Ulm-Augsburg, and between Ulm and Esslingen, extending to the Palatinate and down the Rhine. The map also demonstrates the effect of the linguistic-ethnic boundary that separates the Franconian and Alemannic territories. Although there are some connecting links crossing this boundary, the map suggests fairly independent regions of influence. That is to say, a Franconian-Lutheran region to the north, and a Zwinglian-Alemannic region to the south. The Kraichgau-Heilbronn area is largely "endogamous," in that places of birth and activity of several preachers lie next to each other.

---

[1]"Residency" is a relative or subjective term. In our case it not only refers to the same locality, but its area of influence as well. Persons who were born in a village that lay in the zone of influence of a city should actually be considered as residents, since they often received their schooling there.

Some localities were conspicuous for turning out a number of reformers who were about the same age. Munderkingen,[1] Schlettstadt/Alsace,[2] Stein/Rhine,[3] Bregenz,[4] Constance,[5] and Ettlingen[6] are cases in point. No doubt, the reformers from these towns influenced each other in a significant degree since they were friends from early school years. They often attended the same university later on, and, once dispersed, drew sustenance from each other's beliefs and personal successes. They frequently had a strong impact on the course of the Reformation even though they had no direct, active hand in it.

In localities where an already established Protestant preacher came from the outside, we often notice a pronounced interest by rulers and/or local citizens in the new ideology. These towns and villages often had close contact with a Reformation center nearby. The knightly territories fall into this category. The origin of the preachers in the mining communities is uncertain, although it would seem that they, too, were hired from outside. Saxon or Tyrolean miners would hardly have been content with a preacher who had a different cultural and linguistic background from their own. Besides, these small villages in a remote mountain region could not have furnished their own Lutheran parsons, particularly at such an early date.

Informal Diffusion

Throughout the Southwest, an awareness of the new teaching among the intelligentsia and the populace usually preceded the presence of a preacher. He either came on request of a certain group from another city or town, or was resident of the community. The preacher was influenced from two directions, namely connections with a person or persons in another city who had already adopted the new religion or links with a local group which likewise favored the new cause. In Freiburg, Ulm, and Rottweil, where no professional advisor was available in the first few years, the movement centered around persons of

---

[1] Paul Beck, Conrad Sam, Johannes Wanner, and Urban Unger.

[2] Martin Bucer, Achilles Gasser, Paul Phrygio, Beatus Rhenanus, and Johannes Sapidus.

[3] Johannes Piscatorius, Erasmus Schmid, and Leonhard Wirth-Hospinian.

[4] Thomas Gassner, Siegmund Rötlin, Johannes Möck, and Jakob Grötsch.

[5] Ambrosius and Thomas Blarer, Johannes Zwick.

[6] Matthäus Erb, Caspar Hedio, and Franciscus Irenicus.

high intellectual calibre who either preached or acted as opinion leaders. The absence of an official preacher did not mean the absence of an evangelically-oriented group, however. In Ravensburg, the humanist Michael Hummelberg, Matthäus Uelin, Joachim Egellius, and a few others formed a tightly-knit Protestant interest group.[1] But none of these persons felt called upon to act as preacher outside their circle, as the majority of the Ravensburgers and their magistracy did not desire any religious change. The decisive second step in an essentially two-step flow of communication, namely the effect of opinion leaders on the population following their interpersonal persuasion, did not materialize in Ravensburg. Another example of this was the town of Marbach near Stuttgart.

Travelling salesmen and journeymen, who distributed religious propaganda materials were often the first to familiarize citizens with Luther's ideas. Yet, the writings of Luther and his disciples affected only a small section of the total population. Since a large percentage was illiterate, it is not surprising that this means of communication was rather restricted in scope. Thus, at least in the beginning, most were informed by word of mouth.

Luther's works appeared in Constance in October, 1518 and in Strasbourg and Memmingen in 1519, while in Biberach booksellers from nearby Memmingen sold his writings in 1521. In the imperial cities of Esslingen and Reutlingen, Lutheran literature appeared in 1520; in Bopfingen they are known to have been distributed as early as the end of 1517. Probably they came from neighboring Nördlingen which hosted an annual fair that attracted businessmen from all over southern Germany. Schaffhausen, Stein/Rhine, and Überlingen experienced the first wave of literature, probably arriving from Constance in 1520. By 1523, the territory of Württemberg was so saturated with Lutheran propaganda, that on November 26, 1522, the Austrian government issued a decree that deplored the many "errors and heresies . . . of Martin Luther . . . written, printed and distributed in Latin and German."[2]

The exodus from the monasteries, especially Luther's Augustinian order, began about the same time. Beginning in 1522, the Augustinians in Mindelheim left the monastery to look for jobs in the town and further away. Four years

---

[1] Ravensburg students in Wittenberg outnumbered those from Ulm. Hummelberg showed great respect for Zwingli, and had good connections with Wittenberg thanks to his friend Thomas Blarer. He forwarded Zwinglian literature to Luther, and reverse. See Gottfried Holzer, "Der Streit der Konfessionen in der Reichsstadt Ravensburg" (unpublished Ph.D. dissertation, University of Tübingen, 1950), p. 20.

[2] Hoffman, "Reformation im Bezirk Besigheim," p. 146.

later, the institution had ceased to exist.[1] This egress may have been attributable to Johannes Wanner, who studied with Luther, and preached here a short time after leaving his native Kaufbeuren. He soon went to Constance. The habit of leaving the cloisters, rarely practiced before Luther's appearance, became widespread in the 1520's. For example, many monks, Michael Stiefel and Martin Fuchs among them, left the Augustinian order in Esslingen and became active spokesmen for the Lutheran cause.

Localities with Resident Preachers

A good number of evangelical preachers already resided in the place of their subsequent activity. Not surprisingly, they often appeared in places that experienced a political reformation from above; here, the territorial ruler--a prince, duke, or, in the case of cities, the magistracy--enforced the new faith. In many instances, the conversion of preachers was therefore not the result of a sincere change of mind, but rather a move to hold on to their job. Some made a last-ditch effort to continue the old faith, while outwardly professing the ideas of local reformers or those hired from other cities. Still, the low supply of Protestant preachers at the turn of the third decade forced authorities to rely on every available person.

In other places, the resident preachers had close ties to former student colleagues who had switched to the new teaching. Sometimes they enjoyed a good relationship with a preacher in an adjacent town. Like the non-resident preacher, the resident preacher's office was contingent on the collective mind of the parish; but unlike the former, he could gradually prepare his congregation for the change.

The Territories of Ulm and Biberach

In the territory of the Free City of Ulm, the citizens of larger towns like Geislingen and Langenau showed sufficient interest in the new movement to make proselytizing unnecessary. It was different in the smaller settlements, where approximately half of the preachers shifted from the old to the new faith while retaining their jobs. For example, in Stötten, the Catholic priest continued in his office for almost a year until the magistracy in Ulm replaced him

---

[1] Andreas Haisch, Der Landkreis Mindelheim in Vergangenheit und Gegenwart (Mindelheim: Kreistag und Landratsamt, 1968), p. 340.

with a Protestant preacher.[1] In the same vein, the Catholic chaplain of Stubersheim, Johann Mann, became the first evangelical preacher until superseded by a Biberach colleague a few months later.[2] In Weidenstetten, the residing parson was allowed to stay in office by acknowledging the Christianity of the 18 Reformation Articles of Ulm.[3] Again, in Altenstadt, Hans Russ and Conis Krapf outwardly confessed to the Articles and so kept their positions as parson and chaplain, respectively. In their thinking, however, they remained strictly Roman-Catholic and were replaced after a probation period in early 1532.[4] Similarly, when they were asked to destroy their altars and images by the officials of the city of Ulm, the inhabitants of Altenstadt refused to follow orders.[5] In Bernstadt, Bräunisheim, Ettlenschiess, and Hofstett-Emerbusch, the Catholic clergy also switched to the evangelical faith, although in a superficial spirit.

Some of the villages that belonged to the city-state of Biberach were also reformed in 1531. The Catholic parson was usually ousted and replaced by a Protestant preacher from Biberach. Oberholzheim was an exception. Here, the priest was partial to the Protestant religion and was discharged in 1527 by the ecclesiastical court in Constance on grounds of "derogatory and disobedient" behavior. He was forced to accept the new creed in 1536 on demand of the Biberach magistracy.[6]

The Danube Towns

The towns of Riedlingen, Leipheim, Lauingen, and Dillingen on the upper Danube also had indigenous preachers.[7] As early as 1520, the 51-year old preacher of Riedlingen, Hans Feylmeyer, began to proclaim Lutheran ideas. Where he got these ideas is not known, but one can assume that they arrived in

---

[1] Christian Sigel, Das evangelische Württemberg: Seine Kirchenstellen und Geistlichen von der Reformation an bis auf die Gegenwart (Gebersheim: typed copy in the Württembergische Landesbibliothek, 1938), Vol. VII, Pt. 1, p. 147.

[2] Ibid., Vol. VII, Pt. 1, pp. 426-27.   [3] Ibid., Vol. VIII, Pt. 2, pp. 696-97.

[4] Ibid., Vol. II, Pt. 1, p. 93.

[5] "Württembergisches aus römischen Nuntiaturberichten 1521-32," Blätter für Württembergische Kirchengeschichte, VIII (1893), 79.

[6] Sigel, Das evangelische Württemberg, Vol. V, Pt. 2, p. 959.

[7] Hermann Clauss, Die Einführung der Reformation in Schwabach 1521-1530, Quellen und Forschungen zur bayerischen Kirchengeschichte, II (Leipzig: Heinsius, 1917), 60.

the area via travelling merchants and salesmen from Switzerland and the vicinity of Strasbourg. A large load of the freight that moved through the town consisted of wine that was shipped from Alsace and Baden to Ulm and Bavaria, while salt was being shipped in the opposite direction. A number of Riedlingen citizens had a share in this trade.[1]

Johannes Zwick, a patrician from Constance, took over the benefice in Riedlingen in September or October 1522, thus bolstering the cause of the growing Protestant community surrounding Feylmeyer. The latter served as the primary target of the lower clergy because of his radical beliefs; and, contrary to Zwick, he could be recalled if his performance should prove unsatisfactory.

Beginning in the summer of 1523, Hans Jakob Wehe conducted Lutheran church services in Leipheim on the Danube below Ulm. At the same time his cousin Eberlin of Günzburg preached in the town of the same name[2] three miles downstream from Leipheim. Wehe certainly was influenced by his fiery reform-minded cousin and, like him, refused to switch to the Zwinglian doctrine practiced by neighboring Ulm.[3] While Eberlin soon was expelled from Hapsburg-ruled Günzburg, his cousin actively continued in his reformatory capacity, serving the communion in both forms as well as abolishing mass in 1524.[4] Günzburg's and even Ulm's citizens attended Wehe's services in Leipheim.[5] In spite of the protest launched by Bishop Stadion of Augsburg against Ulm's condonement of Wehe's hereticism, the magistracy there vowed to support Wehe. Ulm was forced to prohibit further sermons by Wehe, but he continued to preach in the Ulm environs before returning to Leipheim early in September, 1524. Notwithstanding the pleas of the Leipheimers, the Ulm magistracy upheld the expulsory order, which prompted Wehe's followers to side with the rebelling peasants in 1525.[6]

---

[1] Oberamtsbeschreibung Riedlingen (2nd ed.; Stuttgart: Kohlhammer, 1923), p. 428.

[2] Paul Auer, Geschichte der Stadt Günzburg (Günzburg: Donau-verlag, 1963), p. 49.

[3] Hans Hopf, "Jacob Wehe: Erster lutheranischer Pfarrer in Leipheim," Beiträge zur Bayerischen Kirchengeschichte, III (1897), 147.

[4] Zoepfl, Das Bistum Augsburg, pp. 40-41.

[5] Hopf, "Jakob Wehe," p. 147.

[6] Zoepfl, Das Bistum Augsburg, I, 41. It seems doubtful that Wehe incited the peasants to rebel against their authorities. See Hopf, "Jakob Wehe," p. 156.

The towns of Lauingen and Dillingen, some eleven and thirteen miles east of Günzburg, lay in the Duchy of Neuburg and Bishopric Augsburg, respectively. Caspar Amann, a prior of the Augustinian Eremite monastery in Lauingen, was a friend of Veit Bild, the humanist and monk at St. Ulrich's in Augsburg. Beginning in August, 1521, the latter obtained Greek works and dictionaries as well as theological titles for Amann. Bild, an early enthusiast of the Reformation, in a letter written in summer 1519, asked Oecolompadius to sell him Luther's "Sermon." It restated the 95 Theses, which he added to his collection of Luther's writings.[1]

Amann also became a good friend to the younger Caspar Haslach, the foundation preacher of Dillingen and a Wittenberg graduate of 1505, before Luther's appointment to the theological faculty.[2] It seems that Haslach became interested in Luther's teaching by way of this friendship. Amann, one of the greatest scholars of that time, was very competent in Greek as well as Hebrew. He gave lessons to Haslach in these languages and afterwards took the opportunity to discuss the burning questions of religious renewal.[3] Veit Bild also began to correspond with Haslach in March, 1522. In one of his letters he expressed astonishment that Haslach had failed to pay him a visit while on stop-over in Augsburg. Bild finally offered the "evangelicae doctrinae concinnator" (advocate of the evangelical doctrine) in Dillingen to his friend.[4]

As in many other places, we here notice a bond between humanist religious interests and a desire for ecclesiastical alternatives. Humanism, as presented by these persons, not only denoted a literary movement but also a belief that the contrast between laity and clergy was outdated.[5]

Veit Bild and the Adelmann brothers rejected the scholastic notion that theology is a field reserved for abstract speculation. Instead, they emphasized the re-invention and glorification of the Bible, a feature stressed by Luther himself. To a large degree, these humanists represented and shaped public opinion and it was they who carried Luther's ideas from the small frontier town of Wittenberg to reformers and congregations in southwestern Germany.[6]

---

[1] Ludwig Duncker, "Die Stellung des Prädikanten Kaspar Haslach zur Reformation," Zeitschrift für Bayerische Kirchengeschichte, XIV (1939), 134.

[2] Ibid., p. 130.   [3] Ibid., p. 134.

[4] Schröder, "Der Humanist Viet Bild, Mönch bei St. Ulrich. Sein Leben und sein Briefwechsel," p. 209.

[5] Kaufmann, Geschichte der deutschen Universitäten, II, 522.

[6] Moeller, "Die deutschen Humanisten und die Anfänge der Reformation," p. 283.

Luther's first appearance in Augsburg greatly stimulated Bild's interest in theological problems. In fact, the entire humanistic community received Luther's writings with alacrity. During the first quarter of 1518, Bild received two of the sermons as well as news of subsequent publications from Bernhard Adelmann, the Augsburg canon. Spalatin, too, kept Bild informed about the newest developments in Wittenberg. He was eager to correspond with Luther. Bild's enthusiasm came to a sudden halt with the peasants' rebellion, since it interrupted the harmonious development of knowledge and challenged the social and political status quo. Bild now complained about the low assessment given to intellectual achievement, and the noticeable drop in school attendance. It looked as if humanism was beginning to be replaced by the Reformation.[1]

Constance

Constance was one of the few upper German cities that had a Protestant preacher as early as 1519. Its convenient location along the Swiss-German border made it a natural gateway in both directions: into Switzerland to the south, and into southwestern Germany to the north.[2] Moreover, Constance served as a focal point for the entire area between the Alps and the cities of Augsburg, Ulm, Stuttgart, Strasbourg, Basle, and Berne. It also enjoyed easy access to the Alpine passes leading to the Mediterranean region and the Rhine and Danube waterways, leading to densely settled, highly developed manufacturing centers. Although the rise of such seaports as Bruges, Bremen, Hamburg, and Lübeck at the end of the fifteenth century brought on a decline of Constance's role as continental trading and transportation center, it continued its leading role as entrepôt for the linen industries of Ulm, Augsburg, Biberach, Ravensburg, Memmingen, and other cities that depended heavily on exports to the Mediterranean.[3]

Paralleling its strategic position between Switzerland and Germany was Constance's importance as a diffusion center for the Lutheran and Zwinglian doctrines. Except perhaps for university towns, no other southwest-German

---

[1] Schröder, "Der Humanist Veit Bild," p. 188.

[2] Otto Feger, "Konstanz am Vorabend der Reformation," in Der Konstanzer Reformator Ambrosius Blarer, 1492-1564, ed. by Bernd Möller (Konstanz: Thorbecke, 1964), p. 41.

[3] F. J. Mone, "Zur Handelsgeschichte der Städte am Bodensee vom 13. bis 16. Jahrhundert," Zeitschrift für die Geschichte des Oberrheins, IV (1853), 6.

urban center had as striking an impact on the course of the Reformation. Protestants in other areas contacted their fellow-believers in Switzerland via Constance, whose mediating role became evident as the Protestant communities on both sides of the lake sought each other's advice and support. It shared this role with Strasbourg, the metropole in the West that gradually was to take over the position of mediator between the Swiss and the Germans.[1]

As in other cities, popular awareness of Luther's ideas, coupled with the appearance of his writings preceded the public proclamation of his words. With the outbreak of the Bubonic Plague in 1519, the population of Constance suddenly displayed an active interest in the new faith, leading to the wide distribution of many of Luther's books, articles, and pamphlets. The literature originated in St. Gallen and Basle, where students of Vadian at the university in Vienna had introduced it earlier.[2] A great number of laymen were able to read and write, thus enabling them to get acquainted with some of Luther's thoughts.

In this context, Jakob Windner began to preach against the old order in 1519. As a graduate of Basle University and member of the Weavers Guild, he sharply attacked the indulgence practices. He had studied under Thomas Wyttenbach in Basle, the humanist who had such a profound influence on Zwingli. After Windner served as Wyttenbach's helper in Biel, Switzerland, he became the foundation preacher at Constance's St. Stephen church, and afterward was parson at St. John's, the largest parish church in the city.[3] He soon received assistance from Bartholomäus Metzler, who arrived from Wasserburg near Lindau across the lake. Both Windner and Metzler were passionate, devoted preachers whose sermons started a large-scale movement away from the old church. They were joined by Johannes Wanner, who arrived in 1521 from Kaufbeuren.

The early Reformation movement in Constance was to a large degree influenced by Zwingli's initial success in nearby Zurich. The government of that city showed a genuine interest in the propagation of the new ideas. The proximity of the Swiss reformer including the rapid and efficient spread of his teach-

---

[1] Christoph Rublack, "Die Konstanzer Reformation," in Der Konstanzer Reformator Ambrosius Blarer, 1492-1564, ed. by Bernd Möller; and Marte, Die Auswärtige Politik der Reichsstadt Lindau von 1530 bis 1532, p. 21.

[2] Hermann Buck, Die Anfänge der Konstanzer Reformationsprozesse, Österreich, Eidgenossenschaft und Schmalkaldischer Bund, 1510/22-1531, Schriften zur Kirchen- und Rechtsgeschichte, No. 29-31 (Tübingen: Osiander, 1964), pp. 38-39.

[3] Ibid., p. 252.

ings explains the southward orientation of Constance. In the beginning, there were no Protestant cities north of Constance that could serve as diffusion centers. The few that did record a somewhat advanced movement were dispersed and far away.

Lindau

For the Upper Swabian cities, a pro-reform course meant an orientation towards Constance. Lindau was one of the imperial towns that enjoyed especially close ties with that city, since it was the second-most important entrepôt on the lake, shipping corn, salt, linen, and cloth to Switzerland, the Mediterranean, and the upper Rhine. The export of wine from Lindau's immediate hinterland was an important source of income until 1530. Regular grain shipments to St. Gallen and Vorarlberg also came via Lindau, and were unloaded in Rorschach and Bregenz. The latter sent two ships to Lindau every Saturday to collect the grain and distribute it in its hinterland;[1] in return, the mountain peasants sold their dairy products there. It is not surprising, therefore, that Lindau served as a refuge and stepping-stone for a number of reformers from the Allgäu cities of Bregenz, Feldkirchen, and Bludenz.

Michael Hugo, a monk, served as the first Protestant preacher of Lindau and after 1522 as <u>Lesemeister</u> (lecturer and preacher) at the local Franciscan monastery. He was the first to give Lutheran sermons in 1522. It is difficult to know how Hugo was converted; one author suggests that the Lindau humanist Achilles Gasser brought Luther's writings to him.[2] Gasser apparently undertook this mission at the request of Urbanus Rhegius, who at that time was residing in his hometown Langenargen, near Lindau. Rhegius had been excommunicated from Augsburg for spreading "Lutheran heresy." As Gasser's private tutor in physics, he managed to convert the young man from Schlettstadt/Alsace. Gasser entered the University of Wittenberg in 1522, inducing Michael Hugo to become a passionate preacher of the new faith. But Hugo, who was described

---

[1] Hans-Gerd von Rundstedt, Die Regelung des Getreidehandels in den Städten Südwestdeutschlands und der deutschen Schweiz im späteren Mittelalter, Beiheft zur Vierteljahrschrift für Sozial- und Wirtschaftsgeschichte, No. 19 (Stuttgart: Kohlhammer, 1930), p. 37.

[2] Albert Schulze, Bekenntnisbildung und Politik Lindaus im Zeitalter der Reformation, Einzelarbeiten aus der Kirchengeschichte Bayerns, Fotodruckreihe No. 3 (Nürnberg: Verein für Bayerische Kirchengeschichte, 1971), p. 5.

as a very pious man, died of the pest in September, 1524.[1]

On Rhegius' request, Gasser delivered Luther's writings and other Reformation literature to Siegmund Rötlin, vicar of St. Stephen, the city church of Lindau.[2] In spite of Rötlin's appointment by the staunchly anti-Lutheran vicar general Johann Fabri of Constance, he gravitated toward the new faith. Rötlin, like Hugo, was a Franciscan monk, and the two strived to assist each other whenever possible. Hugo's determined stand for the Protestant cause made a convert out of Rötlin.

Rötlin's close friendship with Zwingli earned him the title "trumpet and horn of Zwinglianism." In this role he was supported by Rhegius, who used his prestige to influence the citizens and magistracy of Lindau.[3] It was not until 1524, however, that they gave their support to the Lutheran cause. Contrary to Fabri's order to discharge Rötlin, the magistracy followed the wish of the people and kept him in office until his death in November, 1525.[4]

After Hugo's death in autumn, 1524, Thomas Gassner was hired from Bludenz, Austrian Vorarlberg, where he had escaped prison on charges of Lutheran heresy.[5] He had a close working relationship with Rötlin and came to believe in Zwingli's symbolic interpretation of the Lord's Supper, which was served in Lindau for the first time in December, 1525. With Rötlin's death, Gassner became city parson and Lindau's chief reformer, a position he kept until the 1540's. He soon backed off from the Zwinglian interpretation of communion and took a stand between Luther and Zwingli. Gassner had been a student in Wittenberg,[6] but was too close to Switzerland to embrace fully the Lutheran position.

In the parishes of Schachen, Eriskirch, and Laimnau, situated along the

---

[1] Karl H. Burmeister, Thomas Gassner: Ein Beitrag zur Geschichte der Reformation und des Humanismus in Lindau (Lindau: Museumsverein, 1971), p. 20.

[2] Karl H. Burmeister, Achilles Pirmin Gasser. 1505-1577: Arzt und Naturforscher, Historiker und Humanist (Wiesbaden: Steiner, 1970), I, 18.

[3] A. Reinwald, "Das Barfüsserkloster und die Stadtbibliothek in Lindau," Schriften des Vereins für Geschichte des Bodensees, XVI (1887), 147.

[4] Schulze, Bekenntnisbildung und Politik Lindaus im Zeitalter der Reformation, p. 13.

[5] Burmeister, Thomas Gassner: Ein Beitrag zur Geschichte der Reformation und des Humanismus in Lindau, p. 19.

[6] Ibid., p. 16.

lake northwest of Lindau, the Reformation likewise was introduced by local residents. In all probability they had contacts with the reformers in Lindau or the humanist Urbanus Rhegius of Langenargen during 1522-1524. Some clergymen of the communities around Langenargen and Lindau frequently visited Rhegius, although the majority of them aroused his dissatisfaction because of their subservient attitude toward the Pope, which included full support of indulgence and mass.[1] Through the agency of Sigismund Röbli, Rhegius resumed contact with Zwingli.[2] At first he shared Zwingli's views about communion, but later came to adopt a middle position between that of the Zurich reformer and Luther.

Dr. Philipp Melhofer of Eriskirch and Christian Herbstmayer of Schachen pooled their energy and resources to launch an attack against the traditional church. In 1525, Melhofer, then either minister or chaplain, published a paper in Augsburg that condemned the mass as a deplorable disgrace to the sufferings of Christ. Herbstmayer, of Fischbach near Buchhorn, was minister in Schachen.[3]

Nearby Wasserburg was the home of the Constance preacher Bartholomäus Metzler. Where he was converted is not known, but it is fair to assume that the university in Freiburg played an important part, especially in light of the fact that this tiny community registered not less than nine students there between 1518 and 1524.[4]

In Laimnau's St. Peter, which was under the patronage of the hospital in Lindau, the priest Oswald Egg and his chaplain are said to have defected from the Catholic Church in 1527. It seems that Egg had personal connections with Thomas Gassner in Lindau. He was forced to spend the night prior to his wedding (August, 1534) in Lindau, because Monfort officials made threats on his life.[5] Once the Reformation had won the upper hand in this town, the magistracy began to center attention on the hinterland and its villages. The latter

---

[1] Gerhard Uhlhorn, Urbanus Rhegius: Leben und ausgewählte Schriften (Elberfeld: Friderichs, 1861), p. 46.

[2] Ibid., p. 47.

[3] P. Beck, "Die Reformation in Riedlingen und ihr Herold," Württembergisches Vierteljahrheft für Landesgeschichte, IV (1895), 172, footn. 1, and Gustav Bossert, "Ein unbekannter Volksschriftsteller der Reformation," Zeitschrift für kirchliches Wissen und kirchliches Leben, V (1884), 423-40.

[4] Mayer, ed., Matrikel der Universität Freiburg von 1460 bis 1656, I, 215-30.

[5] "Prediger-Historie der Reichsstadt Lindau im sechzehnten Jahrhundert," Historisch-Politische Blätter, LXII (1868), 526.

were incorporated into St. Stephen, where the city or its hospital had the collation right to designate the priest. The preachers who came here in 1528 and 1529 began to reform these communities. The ius patronatus of the Lindau hospital in addition to Laimnau also covered the villages of Reutin, Weissenberg, and Lindenberg.[1] Here, the task of initiating the people was relegated to the Constance reformer Johann Zwick, in the years 1528 and 1529; but he was repulsed by peasants armed with flails and forks as he tried to enter their villages.[2] This is an illustration of the conservatism of the population and clergy in Lindau's outskirts that Rhegius deplored so much.

Isny

This imperial town between Lindau and Kempten enjoyed a lively trade with the rest of the imperial cities of Upper Germany. In addition, kinship ties with Nuremberg, Kempten, Memmingen, and St. Gallen probably favored the influx of reformatory ideas into the area. For example, Peter Buffler, the local merchant, regularly received literature and news about the movement from his brother in Nuremberg. As soon as he read it, he passed it on to the foundation preacher Conrad Frick and his aides. They, in turn, used such literature as the basis for their sermons.[3]

Cities on the Upper Rhine

The upper Rhine between Constance and Waldshut formed a cross-road between the Lake of Constance and Strasbourg and between Switzerland and the interior of Swabia. Whereas Constance specialized on long-distance trade between Italy and southern Germany, Schaffhausen functioned as a traffic depot for the shorter trade routes between Switzerland and Württemberg. Stein and Diessenhofen upstream from Schaffhausen also were marketing centers along this route. The intensification of communication in the late fifteenth century led to a growing influx of southwest German artisans into Swiss cities. Most went to Zurich, Basle, Berne, and Lucerne, but smaller cities and towns such as Schaffhausen, Winterthur, Baden, St. Gallen, and Stein also experienced

---

[1] Manfred Ott, Lindau. Historischer Atlas von Bayern, Teil Schwaben (München: Müller, 1968), pp. 130-31.

[2] "Prediger-Historie der Reichsstadt Lindau," p. 526.

[3] Immanuel Kammerer, "Die Reformation in Isny," Blätter für Württembergische Kirchengeschichte, LIV (1954), p. 4.

this influx. The ties these artisans felt between their adopted cities and their former hometowns proved to be significant in the Reformation movement of the 1520's. First, many of the "Swiss" reformers were actually Germans from the Southwest who followed their comrades south when authorities prohibited the preaching of Lutheran sermons. When the Duchy of Württemberg experienced the political reformation beginning in 1534, many of them returned north, accompanied by native Swiss preachers, where they reoccupied newly-opened positions in the Alemannic portion of Württemberg, south of Stuttgart. Secondly, many of the earliest religious exiles, such as those from Rottweil, settled in the northern Swiss cities. Together with Constance and Berne, these cities absorbed no less than a third of Rottweil's expellees.[1] Indeed, Schaffhausen, Diessenhofen, and Stein each absorbed twice the number of people who settled in Constance.

In Stein/Rhine, a Benedictine monk from Constance, Erasmus Schmid, hailed Luther's religion as "returning Christianity." Since Schmid, a former student at Freiburg/Breisgau and Basle had exchanged letters with Zwingli as early as June 12, 1518, it is safe to assume that the Zurich reformer was instrumental in his conversion. The resulting friendship between the two was encouraged by Hans Öchslin, a friend of Schmid and the parson in Burg across the river from Stein.[2]

The Protestant community in Stein had at their disposal two additional organizers of the movement, Leonhard Wirth-Hospinian, and Johannes Piscatorius. Wirth was a fellow-student of Thomas Blarer and Conrad Zwick in Wittenberg. His father was an innkeeper, who must have heard many discussions and arguments concerning the new religion. Piscatorius was converted early by the humanist and doctor Wolfgang Rychard during his stay in an Ulm monastery. Because of Stein's excellent business connections with cities along Lake Constance and beyond, Wirth, Schmid, and Piscatorius were able to promote Zwingli's teaching in Lindau and the Allgäu.[3]

Thanks to Erasmus Schmid, the clergy in Schaffhausen in 1520 learned to

---

[1] Heinrich Ruckgaber, Geschichte der Frei- und Reichsstadt Rottweil (Rottweil: Englerth, 1894), pp. 244-45.

[2] Jakob Wipf, Reformationsgeschichte der Stadt und Landschaft Schaffhausen (Zürich: Füssli, 1929), p. 74.

[3] Hildegard Urner-Astholz, et al., Geschichte der Stadt Stein am Rhein (Bern: Haupt, 1957), pp. 158-59; and Buck, Die Anfänge der Konstanzer Reformationsprozesse, p. 196.

familiarize themselves with Luther's works. The reformer of Schaffhausen, Sebastian Wagner (alias Hofmeister) was a native there. A Franciscan friar, he became lecturer of the Zurich chapter, and it was in this capacity that he came to know Zwingli.[1] He earned a doctorate of the Holy Scripture, the highest academic degree, at the University of Paris in 1519-1520--at a time when Luther's concepts rocked the intelligentsia there. The accompanying religious persecution forced Hofmeister to return to Zurich where he became lecturer and a close associate of Huldrych Zwingli for the rest of his life. Shortly afterward, his superior sent Hofmeister to Constance, probably to avert any further contact with Zwingli and other converts. Hofmeister carried on his correspondence with Zwingli and Luther, however, and he envisioned a new world order based on the realization of Luther's ideas.[2]

Zwingli himself was at that time indebted to Wittenberg for his religious ideas. Not until 1521 did he develop his own version of Protestantism;[3] his distinctly personal point of view concerning the communion did not become popular until several years later. With great courage Hofmeister began his Protestant worship in the middle of 1522, addressing every segment of the Schaffhausen community. The preparatory groundwork had already been laid by a group gathered around the last abbot, Michael Eggenstorfer of All Saints, Erasmus Schmid, Zwingli's friend, the Protestant city physician Johann Adelphis, and Ludwig Öchsli, the schoolmaster.[4] However, as early as 1522, Hofmeister approved Zwingli's concept of communion as a symbolic act, and thus rejected Luther's notion of the real presence of Christ's body and blood.[5] In a letter dated one week after Easter 1523, Hofmeister placed his entire hope for the sustenance of the evangelical movement on Zwingli's rather than on Luther's stand against the bishops of Constance and Veroli.[6]

In August, 1522, after numerous quarrels between Catholics and Protestants, the Schaffhausen magistracy intervened in favor of the new party and demanded that all preachers use the Bible as their sole guideline. Yet, the decision reached by the Diet in Nuremberg emphasized the status quo in the reli-

---

[1] Wipf, Reformationsgeschichte der Stadt und Landschaft Schaffhausen, p. 68.

[2] Ibid., p. 113.

[3] W. Köhler, Zwingli und Luther: Ihr Streit über das Abendmahl nach seinen politischen und religiösen Beziehungen (Leipzig: Heinsius, 1924), I, 256.

[4] Wipf, Reformationsgeschichte der Stadt und Landschaft Schaffhausen, p. 123.

[5] Ibid., p. 127.   [6] Ibid., p. 132.

gious contest. And since Schaffhausen was more open to attack from the north than the Swiss territory south of the Rhine, it adopted a more reserved or respectful attitude toward the Empire and its emperor.[1]

In this climate, the Reformation once more gained adherents. In November, 1524, images were destroyed by three citizens who were exiled for two years. This is a clear indication that the movement had become distinctly "Swiss" in character, although Austrian officials in late 1524 still labeled it as "Lutheran."[2] By this time, most of the inhabitants and clergy had abandoned the old faith.

Waldshut is located halfway between Basle and Schaffhausen, as well as between Basle and Zurich along the busy land and water route of the Rhine. Nearby Zurzach hosted the large annual fair that attracted buyers and sellers from all over the Southwest, Alsace, and Switzerland. That some inkling of Luther's ideas existed in this area is demonstrated by a student who attended Wittenberg in 1521.[3] However, most students preferred the universities of Basle and Freiburg. The former functioned as a grain supply center for the towns of Waldshut, Laufenburg, Rheinfelden, and Baden. Corn dealers from these towns obtained their supplies in Basle and sold them either on the local market or further inland in Zurich and Lucerne.[4] Forthwith these larger cities were to play a significant role in introducing and consolidating the Reformation in Waldshut and Rheinfelden.

It was the widely travelled Dr. Balthasar Hubmaier who was responsible for the introduction and implementation of the new faith in Waldshut. He was a graduate of Freiburg, where Zell, Sturm, Rötlin, Ofner, Wilhelm of Fürstenberg, Johann Zwick, and Oeler were also students. All were to become converts and local reformers in the upper Rhine region between Strasbourg and Lindau. Hubmaier arrived there in early 1522. Forced out of his position in the Regensburg cathedral due to some favorable comment he made concerning Luther, he became a school teacher in St. Gallen. Shortly afterward, the Swiss convent transferred him to a church in Waldshut where he preached in the old tradition.[5] Although Hubmaier did possess the highest degree academically,

---

[1] Ibid., p. 134. [2] Ibid., p. 155.

[3] Christian Connersmannes, October 10, 1521.

[4] Rundstedt, Die Regelung des Getreidehandels in den Städten Südwestdeutschlands, p. 34.

[5] Vierordt, Geschichte der Reformation im Grossherzogtum Baden, I, 192.

he was not able to reach and enforce his own opinion about Luther. He relied on discussions and advice from Erasmus, Glarean, and others in Basle whom he visited freely, as well as the reformers in Schaffhausen and Zurich: Zwingli, Oecolompadius, and Vadian. His assiduity and eloquence in rendering the Gospel soon won him the majority of the population of Waldshut.[1] Also, in 1523, he attended the religious discussion held in Zurich, where he still cautioned against radical changes in ritual that occurred locally under Zwingli's leadership, although he was ready at this time to approve communion in both forms and reading of the mass in German. Hubmaier's evangelical circle in Waldshut continued to grow, and in a meeting on May 14, 1524, its citizens decided to accept the evangelical creed officially. But the Austrian government intervened. Under threat of a military attack, it demanded the extradition of Hubmaier. Although his followers vowed to protect their leader, Hubmaier temporarily left for Schaffhausen.

In early October, a volunteer corps from Zurich also entered Waldshut in order to "protect the Divine Word."[2] Three weeks later Hubmaier was given an enthusiastic reception upon his return. The succeeding reforms included the abolition of mass and images. As these actions resembled the measures taken by Zwingli in Zurich, Hubmaier called himself "brother of Huldrych Zwingli in Christ." He did not stop here, however, but became a follower of the Anabaptist movement which reached Waldshut in March 1525, from its cradle in Zurich.[3] The rebelling peasants in and around Waldshut again followed Hubmaier, by making him their advisor and leader. The end of the Peasants' War signalled the end of the three-year-old Reformation movement in the town. Hubmaier again fled; this time to Zurich, where he was jailed by the local magistracy.[4]

---

[1] Alfred Stern, "Hubmaier, Balthasar," Allgemeine Deutsche Biographie, Vol. 13, p. 265.

[2] Ibid.

[3] For an excellent treatise on the social composition and diffusion of Swiss anabaptists, see Paul Peachey, Die Soziale Herkunft der Schweizer Täufer in der Reformationszeit, Schriftenreihe des Mennonitischen Geschichtsvereins, No. 4 (Karlsruhe: Schneider, 1954).

[4] Allgemeine Deutsche Biographie, Vol. 13, p. 266.

Upper Neckar Cities

In the upper Neckar basin between Tübingen and Villingen, which later became solidly Catholic due to the politically induced Counter-Reformation that followed the Lutheran and Zwinglian movements, the Protestant influence came largely from two directions: the Switzerland/Lake Constance area to the south, and the Freiburg/Strasbourg area to the west. No influence from the Duchy of Württemberg and its imperial city enclaves further north was discernible in this area. For one thing, the evangelical movement in these domains had not progressed to the same extent as it had in the upper Rhine region. The ties between the latter region and the upper Neckar area were numerous and manifold. Many residents of Rottweil, Horb, Rottenburg, and smaller towns received their education at Freiburg university, while artists and artisans were in constant touch with the Strasbourg and Basle metropoles. At the same time, the upper Neckar cities had firm ties with Switzerland, being a relatively fertile agricultural region that sold large amounts of grain to the Swiss. In addition, the various textile products manufactured in the above-named cities were exported chiefly to the South and West.

The subordinate, even restraining role of the university in Tübingen toward the Reformation calls for more detailed comment. An important factor was the lack of faculty members who could profess a strong interest in Luther's ideas. A second factor was the university's location in an area that was subordinate to the Habsburg rulers in Innsbruck. They saw to it that the Reformation movement in adjacent Reutlingen had no spillover effect on the university community. In addition, Melanchthon's early departure meant the loss of a potential religious leader (he left Tübingen in 1519). Even Duke Ulrich's order that forbade the practice of the Catholic faith in 1534 was met with resistance. The turning point came in 1535-1536, when the theologians were forced to relinquish their position; not until much later, however, did the Protestant theological seminary or _Stift_ come to be known as the "heroic light that penetrates deeply into the dark Popish territory" of Austrian Hohenberg.[1] In the 1520's Tubingen's academic community hardly differed from the latter.

In Rottenburg, the capital of the Austrian-ruled county of Hohenberg, Nikolaus Schedlin was the parson at the parish church from 1517 until 1536.[2] He

---

[1] Lecture notes from Prof. Decker-Hauff.

[2] _Beschreibung des Oberamts Rottenburg_ (Stuttgart: Kohlhammer, 1900), II, 62.

was educated at the universities in Erfurt and Freiburg, where he had earned a degree in theology.[1] His only missing distinction was the insignia, characteristic of a doctor of theology.[2]

The second evangelical preacher, Hans Eycher first occupied the canonry, then the preaching foundation at St. Moriz in Rottenburg-Ehingen. His case is discussed later in the chapter dealing with Prädikaturen, or preaching foundations.

Despite the fact that Schedlin later renounced his earlier religious convictions, he provided the initial impetus that set the wheels of the Reformation movement in motion. A report to the Bishop of Constance by the Rottenburg advocate mentions Schedlin as having originated the basis and beginning of the "Lutheran heresy." As the teacher of Hans Eycher,[3] he became the opinion leader in and around Rottenburg, until he was forced out of office in 1527. A son of a Rottenburg citizen, possibly a fisherman, Schedlin received an excellent education at three different universities.[4] The University of Freiburg, which he attended last, also was patron of Rottenburg's parish church.[5] It was through its agency that Schedlin became its parson and a convert to the new faith. His fellow students in Freiburg were Matthias Zell, Jakob Sturm, and Wolfgang Capito. Most probably, Schedlin continued to correspond with these leading figures. Moreover, he offered religious instruction in his own home through a curate[6] from Esslingen who was declared to be a "heresis lutherana

---

[1] Mayer, Die Matrikel der Universität Freiburg i. Br. von 1460 bis 1656, I, 155.

[2] Since the bestowal of these insignia, including the ceremony, required a considerable sum of money, many candidates (Schedlin seems to be included) were content with the licentiate. See Werner Kuhn, Die Studenten der Universität Tübingen zwischen 1477 und 1534: Ihr Studium und ihre spätere Lebensstellung (Göppingen: Kümmerle, 1971), I, 29.

[3] Bossert, "Kleine Beiträge zur Geschichte der Reformation in Württemberg," p. 149.

[4] Sigel, Das evangelische Württemberg, Vol. VI, No. 2, pp. 821-22.

[5] Mayer, Die Matrikel der Universität Freiburg i. Br. von 1460 bis 1656, I, 154.

[6] The accusation in a letter stems from Dr. Balthasar Sattler, a former Tübingen professor, then parson in Canstatt, and seems to have been directed at Johann Lonicer, who was forced to leave Esslingen for Strasbourg. Lonicer was a philologist and professor of Greek and Hebrew in Freiburg. See Horawitz, "Lonicerus, Johannes," Allgemeine Deutsche Biographie, Vol. 19, pp. 158-63.

convictus" by one of its parsons. Schedlin offered the curate shelter under the pretext that the latter offered him lessons in Greek.[1]

In September 1523, Eberlin of Günzburg arrived in Rottenburg directly from Wittenberg. Although his stay was short, his impact was great, since Eberlin was well known in Rottenburg, particularly in the duchy of Hohenberg. There he had helped to strengthen the Catholic cause in the years 1517-1519, shortly prior to his conversion. In July, 1523, he addressed an open letter to the Hohenbergers in which he explained and defended his break with the Catholic faith as a conscientious duty toward the Scripture, and warned the people of false prophets. No doubt, his ardent exertions in Rottenburg added considerable momentum to the movement that was already brought into gear by Schedlin and Eycher.[2]

Schedlin, at first enthusiastic about the Lutheran cause, anxiously withdrew from the scene as soon as he perceived the weight of the opposition of the Austrian government. As a regular preacher, he was bound to come into open conflict with political authorities sooner than Eycher, who had a preaching foundation. While avoiding open conflict with the old church, Schedlin simultaneously tolerated the Protestants in town. He appeared to be a calm, intellectual person, disinterested in religious quarrels and bickering which were so common during that time.[3]

Schedlin's disciple, Johann Hechinger, followed his master and attacked papal indulgence as a fraud. He was captured and hanged during the Peasants' War.[4] Andreas Keller, a former monk, also began to preach the evangelism after obtaining a benefice in Rottenburg in 1524. He also attacked the Pope and proclaimed Luther's thesis of justification by faith, but he was forced out of the town at the end of 1524.[5] The magistracy of Strasbourg obtained a preaching post for Keller in Wasselnheim[6] near Strasbourg. From there he continued to

---

[1] Bossert, "Kleine Beiträge zur Geschichte der Reformation in Württemberg," p. 149.

[2] Sigel, Das evangelische Württemberg, Vol. VI, No. 2, pp. 827-28.

[3] Ibid., p. 823.

[4] Bossert, "Kleine Beiträge zur Geschichte der Reformation in Württemberg," p. 149.

[5] Ibid.

[6] Adam, Evangelische Kirchengeschichte der Elsässischen Territorien bis zur Französischen Revolution, p. 64.

cultivate the new faith in Rottenburg. Keller's affiliations with Strasbourg and Wasselnheim supports our earlier thesis that Schedlin was influenced by his former study companions in Freiburg, namely Zell, Sturm, and Capito. A long and fruitful friendship between these men made it possible for Keller to preach the gospel for twelve full years in the Alsace, before Ambrosius Blarer called him back to Wildberg in 1536.

Middle Neckar Area

Esslingen was an old imperial and wine export city of approximately 8000 inhabitants. It functioned as central place for the surrounding Neckar valley, the Filder Plateau, Schurwald, and the Rems and Fils valleys. In addition, the city acted as export agent for the many wine-growing communities in the Rems valley. It was particularly suited for this role since it was located on the major road from Speyer to Bavaria which transversed Stuttgart, Göppingen, Geislingen, and Ulm. Esslingen also enjoyed firm economic ties with the upper Rhine and the Alsatian cities.[1]

Whereas the initial impulse for religious reform in Esslingen emanated from Strasbourg via Stuttgart, the major impetus came from the larger and highly respected imperial city of Ulm. Esslingen copied not only Ulm's laws and regulations regarding the power of guilds and the sale of foodstuffs; its master builders also went to Ulm to inspect buildings that were considered exemplary.[2] With the suppression of the Reformation in cities and villages surrounding Esslingen, some preachers moved to the Ulm area. Moreover, the switch made by the Esslingen magistracy from a pro-imperial to a pro-Reformation position was the direct result of a similar move in Ulm.[3] And when Ulm shifted from a pro-Zwinglian to a pro-Lutheran stand in the mid-thirties, Esslingen again followed suit.

The literature of the Reformation first made its appearance in the imperial city in 1520, and the initial spokesman for the Protestant cause was Martin Fuchs, the chaplain at the local parish church since 1518. After receiving a

---

[1] Keyser, Württembergisches Städtebuch, p. 72.

[2] Eberhard Naujoks, Obrigkeitsgedanke, Zunftverfassung und Reformation: Studien zur Verfassungsgeschichte von Ulm, Esslingen, und Schwäbisch Gmünd, Veröffentlichungen der Kommission für Geschichtliche Landeskunde in Baden-Württemberg, Reihe B, No. 3 (Stuttgart: Kohlhammer, 1958), p. 35.

[3] Ibid., p. 87.

Bachelor's degree in Tübingen in 1512 as classmate of Ambrosius Blarer, he entered the Augustinian order. Fuchs held his first Lutheran sermons in 1521 and continued them until his expulsion in December, 1524.

Another Augustinian monk, Michael Stifel, offered Luther his full support regarding the indulgence question. He may have attended the meeting of his chapter in Heidelberg in April of 1518.[1] In any case, Stifel, with the support of his colleagues Hieronymus Gandelfinger, Johann Lonicer, and St. Schäffer, was the leader of the early Protestant movement. In sermons and brochures he defended and promulgated Luther's "well-founded teaching" in the city and beyond. Stifel was the son of a burgher, and although not wealthy, was highly thought of in his community. Unlike Fuchs, he did not attend a university but the local Latin School, which was the equivalent of one year at a university.

Stifel wrote a very popular Protestant song, but was forced to leave town in 1522. Yet, this did not prevent his circle of reformed friends from contacting each other, and it was Stifel who created a direct connection between Esslingen and Luther in Wittenberg. During his last weeks in Esslingen, he developed a close friendship with the Augustinian monk and scholar of classics, Johannes Lonicer, who had just fled from his teaching post at Freiburg University. During his short stay in Esslingen, he probably lived in the Augustine monastery and very likely helped to convert Stifel to Protestantism.[2]

The significance of Lonicer on the movement cannot be overemphasized. Born in Eisleben, he attended the universities of Erfurt and Wittenberg, where he became Luther's teaching assistant in 1520. The kinship that Esslingen felt with Luther was therefore not only the work of Stifel, but also Lonicer. And it was Lonicer who continued, albeit for a short while, Stifel's highly popular sermons in Esslingen. From here he was in close touch with the Lutheran community in Stuttgart, particularly its foundation preacher Johann Mantel, also an Augustinian and a Wittenberg graduate. This "diligent and . . . pleasant preacher" in the Augustinian monastery was expelled from Esslingen in 1523. After a short stay in Rottenburg and Reutlingen, he went to Strasbourg, where he was engaged in furthering the Reformation by preaching, writing, and translating Luther's works.

Nevertheless, stout opposition came from the magistracy and patriciate,

---

[1] Joseph E. Hoffmann, "Michael Stifel: Zur Mathematikgeschichte des 16. Jahrhunderts," Esslinger Studien, XIV (1968), 33.

[2] Gustav Bossert, "Zur Geschichte der Reformation in Esslingen 1522 und 1523," Blätter für Württembergische Kirchengeschichte, N.F. VIII (1893), 93.

led by Mayor Hans Holdermann, who had the backing of the Speyer cathedral chapter and the bishop in Constance. The major cause for the repression of the new doctrine lay in the relocation of the imperial court and government from Nuremberg to Esslingen, where it ruled from 1524 until August 1527.[1] In addition, the city served as a meeting place for the pro-Catholic Swabian Federation. Finally, it was forced to take account of its economic dependence on the government of Württemberg in nearby Stuttgart. These were the factors responsible for the vacillating course of the Reformation in Esslingen. Despite pressure of the governments in the imperial and territorial towns contra the evangelical movement, it continued to grow unrelentingly.

A key factor in the fruition of the movement was the common economic pressure of the guilds, which provided support that was too powerful to be suppressed. The guilds stood for the protection of the many small wine growers whose interests were commonly thwarted by the governing oligarchy, which enjoyed the favor of the Swabian Federation and the Habsburg ruler in Stuttgart. Most wine growers were tenants who worked on property belonging to powerful Esslingen landlords. During the 1520's the latter could count on the assistance of the city government in their rivalry with the small wine growers and their guilds. The landlords aimed at regaining full control over their holdings, including the right to lease to anyone, while the growers thought they had a moral claim to the property since they had worked on it for many years. The contradicting socio-economic interests of the two groups became reflected in the ensuing religious cleavage. The landlords sided with the traditional political and religious establishment, whereas the wine growers and their guilds pressed for change that culminated in the removal of all restrictions applying to the new faith, the invitation of Ambrosius Blarer to Esslingen, abolition of the mass, and the removal of images during the last months of the year 1531. The Esslingen government was encouraged by preceding ecclesiastical reforms in Ulm.

In 1524, Margrave Philipp I of Baden-Durlach[2] brought his court and itinerant preacher Franciscus Irenicus to Esslingen, where he held Lutheran sermons in the Augustinian monastery in which Stifel and Lonicer had preached

---

[1] It was precisely the decision to shift the seat of the imperial government from Nuremberg to Esslingen that first triggered the conflict between the religious interest of the population and the political interest of the magistracy. Now the latter assumed a strict anti-Lutheran position. See Louis Reimer, "Reichsregierung und Reformation in Esslingen," Esslinger Studien, XI (1965), 226-40, esp. p. 228.

[2] He was the newly appointed regular substitute of Archduke Ferdinand.

before.[1] The popularity of Irenicus greatly increased church attendance and drastically curtailed offertory at the Catholic parish church.[2] Consequently, the papal legate, Campegio, lodged a complaint against Margrave Philipp and his preacher, but all he was able to achieve was a shift of Irenicus' popular sermons to another hour of the day, so that Catholic and Protestant services would not coincide. Irenicus was bound to become the religious leader of the Lutheran community,[3] which had an effect even on the imperial government, since it would not act to curb his well-attended sermons. Needless to say, Irenicus' high reputation among Esslingen's Protestants became known to Zwingli in Zurich. Another Lutheran preacher besides Irenicus was Conrad Schlupf, called "belzebub . . . and a hundred-percent Lutheran" by his enemies. Schlupf, a Franciscan friar, was the first known clandestine Lutheran preacher in Biberach in 1523. Like other preachers from Upper Swabia and Württemberg, he must have arrived in Esslingen in 1525 or 1526, through the mediation of Ulm.

The personal interest Brenz and Zwingli took in the city of Esslingen in 1525 did much to promote the evangelical movement. After the magistracy hired Ulrich Villinger as preacher in 1526, the Catholic priest was forced to leave town in 1527-28. Then the mayor, Hans Sachs, and especially its syndicus, Johannes Machtolf, became the secular leaders of the Protestant party. But notwithstanding this obvious shift away from the established church, the city government displayed a moderate stance outwardly. No doubt, this was the result of political and economic dependence on the surrounding duchy of Württemberg and solidarity with the imperial government. Still, as the imperial city failed to take a definite stand on the Peasants' War, the pro-reform minded circles were encouraged to demand radical changes in the ecclesiastical conditions in their town.

The attorney Johannes Machtolf, one of Lonicer's "friends and patrons" was an ardent supporter of the Reformation. After attending the universities in Heidelberg and Freiburg, where he acquired a doctor in law, he returned to his hometown of Esslingen to become its syndicus.[4] It was also Machtolf who led

---

[1] Reimer, "Reichsregierung und Reformation in Esslingen," pp. 230-31.

[2] Its parson, Dr. Balthasar Sattler lamented in 1523 that the four offerings brought only 60, instead of the regular 120 to 130 pounds. While the oblations collected by the chaplains dropped from 60 or 70 pounds to mere 18 pounds, the fruit tithe decreased from the usual 24-25 pounds to 2 pounds. See Krebs, Die Protokolle des Speyerer Domkapitels, II, 100.

[3] Kattermann, Die Kirchenpolitik Markgraf Philipp I., p. 18. footn. 59.

[4] Theodor Keim, Reformationsblätter der Reichsstadt Esslingen (Esslingen: Weychard, 1860), p. 24.

the pro-Zwinglian faction in the successful mission which Ambrosius Blarer performed from August to October of 1531.

The decisive factor which led Machtolf to adopt a more moderate form of Zwinglianism was his personal relationship with Ulm's burgomaster Bernhard Besserer. The latter showed his preference for the Swiss Reformation as early as 1525, by ordering his foundation preacher in Ulm to preach in that style.[1]

The religious upheaval in Esslingen left its mark on a surrounding area that was governed largely by the Austrian Habsburgs in Stuttgart. The impact seems to have been strongest in wine-growing areas on the sunny slopes of the Schurwald, the Rems Valley, and the Asperg. The villages of Mettingen, Sulzgries, and Rüdern northwest of Esslingen as well as the towns of Waiblingen, Schorndorf, and Asperg were typical wine-growing communities in the fertile Neckar-Rems valleys.

The village of Weil im Dorf northwest of Stuttgart had connections with the university in Tübingen. According to a report released in 1531, the village population, including its curator Veit Simon, showed a clear preference for the evangelic faith. We can be sure that this was not transmitted from Tübingen; instead, they must have emanated from Esslingen, as the local Protestants felt a mutual kinship with that city. The village parson, Jakob Ringlin held private sermons for the local peasantry. In December, 1531, he became evangelical preacher in Esslingen, the curator having been fired in May, 1533.[2] With the move of their preacher to Esslingen in September 1531, some villagers continued attending his sermons there, returning with the evangelical message in oral and written form.[3]

In Neckargröningen, Johannes Hannemann preached Luther's faith in 1527, seven years prior to its official introduction by Duke Ulrich. He was subjected to various strenuous interferences by political authorities. Where he made contact with the reformatory movement is difficult to establish.[4] The village of Ditzingen faced a similar situation. Here Rudolf Heim, a former Hirsau Bene-

---

[1] M. Ernst, "Bernhard Besserer: Bürgermeister von Ulm (1477-1542)," Zeitschrift für Württembergische Landesgeschichte, V (1941), 83-113.

[2] Johannes Haller, Die Anfänge der Universität Tübingen (Stuttgart: Kohlhammer, 1929), I, 317; and II, 124. Keim, Esslinger Reformationsblätter, p. 71.

[3] Beschreibung des Oberamts Leonberg (Stuttgart: Kohlhammer, 1930), I, 417.

[4] Bossert, "Kleine Beiträge zur Geschichte der Reformation in Württemberg," p. 161.

dictine monk likewise started sermonizing in 1527. Ditzingen's strategic location along the main road drew a good attendance.[1]

Pforzheim, situated on the northern spur of the Black Forest was the most important territorial city in the Margraviate Baden during the Reformation period. Around 1500, this city of four to five thousand residents was the communication center between the Rhine and Württemberg. Although we lack adequate data on its external affairs, Pforzheim was known for its lucrative business relations with Speyer and Heilbronn, and Speyer functioned as the financial center for a number of imperial cities. While Esslingen and other Alemannic cities in the upper Neckar basin showed an intellectual orientation toward Freiburg, Strasbourg, and Zurich, the Franconian city of Pforzheim leaned more toward Heidelberg and Wittenberg. Pforzheim's famous Latin School turned out a number of brilliant reformers, and had between two and three times as many students in Heidelberg than in Freiburg; the ratio increases to four over one if one adds the number of students in Wittenberg. Hence the precedence of Lutheranism in this area.

Pforzheim, like Rottenburg, Ulm, and Biberach accepted the new faith five years after it had been proclaimed by a young monk, Johannes Schwebel. Schwebel was 29 years old when he began to preach the gospel in 1519 at Holy Ghost Hospital. He had become well acquainted with Melanchthon while attending Pforzheim's Latin School.[2] After studies at the universities in Tübingen, Leipzig, and Heidelberg, he entered the monastery. Schwebel preached so fervently that in June, 1522, Margrave Philipp asked him to leave the city and seek refuge with Franz of Sickingen on the Ebernburg. It seems likely that Schwebel was expelled for trying to introduce liturgical changes.[3] Having served under Sickingen and Count Louis II of the Palatinate, Schwebel visited his hometown in 1524 and held a Protestant service in the hospital church. As

---

[1] Ibid., p. 160.

[2] Aloys Stolz, Geschichte der Stadt Pforzheim (Pforzheim: Stadt, 1901), pp. 48-49.

[3] Dr. Veus, Margrave Philipp's chancellor, agreed with Luther's ideas until the year 1520. The latter's attack on the traditional sacraments and ceremonies caused him to withdraw his support.
We have every reason to assume that Schwebel's dismissal must be seen in a larger religious-political context. After all, Schwebel seemed to have close contacts to Wittenberg, where communion sub utraque specie was served six months earlier. See Kattermann, Die Kirchenpolitik Markgraf Philipps I., p. 12.

a reconciliatory move[1] Margrave Philipp returned a large part of Schwebel's property which the latter had once bestowed upon his monastery in Pforzheim. At the same time, Margrave Philipp I conferred the preaching foundation at St. Michael on Johann Unger. Unger, at one time private tutor of Melanchthon, kept in close touch with the events in Wittenberg. For a more extensive coverage of Unger and the Prädikatur he occupied, the reader is referred to Chapter VII.

In Bietigheim, the Austrian government strictly prosecuted adherents of the Lutheran faith. For example, an unknown clergyman was forced to give up his position upon a decree that the clergy must warn their congregations of Luther's "errors and heresies." Similarly, after the peasants' defeat, Hans Baderlin was made to swear to abstain from holding "wantonly Lutheran phantasies and sermons."[2] Another, town clerk Bernhard Widmann was punished and expelled for his pro-peasant stand.

In adjacent Horrheim, the parson Melchior Reich was executed by Aichelin, the imperial executioner.[3] In the other villages around the town of Besigheim such as Schützingen, Gross-Sachsenheim, Grossbottwar, and Beilstein, the first Lutheran sermons were associated with the residing parson who joined the rebelling peasants. See Chapter II for further details.

A notable exception was the village of Ilsfeld. Its first preacher, Johann Gayling attended Wittenberg university from 1513 to 1518, and again in 1520 after studying in Heidelberg and Erfurt.[4] His father must have been a well-to-do wine grower to have permitted his son to study at so many different places.[5] Gayling was one of six students from Ilsfeld who attended Wittenberg university in the decade between 1515 and 1525. Upon his return from Wittenberg, he pre-

---

[1] That this was not an isolated incident, but a change of mind on the part of Philipp is illustrated by the fact that he kept Irenicus as his court preacher and prebendary in Esslingen, despite the vigorous intervention of the papal legate and the Speyer cathedral chapter. See ibid., pp. 20-21.

[2] Hermann Roemer, Geschichte der Stadt Bietigheim an der Enz (Stuttgart: Kohlhammer, 1956), p. 96.

[3] Sigel, Das evangelische Württemberg, Vol. IV, No. 2, p. 448; and Roemer, Geschichte der Stadt Bietigheim an der Enz, p. 96.

[4] Bossert, "Beiträge zur badisch-pfälzischen Reformationsgeschichte," p. 34.

[5] On the tax list (Herdsteuerliste) of the Lauffen prefecture, the two Gayling families mentioned are near the top. Hauptstaatsarchiv Stuttgart, Bestand A 54a, no. 36.

sumably established contact with Conrad Sam, the foundation preacher in adjacent Brackenheim. A letter sent by Luther via Gayling and addressed to Sam indicates that the Wittenberg reformer was aware of the latter's activities.[1] The contact between Gayling and Sam took place some time between 1518 and 1520, and probably in 1520, when Sam became preacher in Brackenheim.

It seems unlikely that Gayling held sermons in Weinsberg from 1520 to 1522, since Schnepf occupied the preaching office at that time. Instead, Gayling apparently preached in his native Ilsfeld before the end of June, 1520. On White Monday, he enrolled at Tübingen university, but for unknown reasons soon returned to Wittenberg. In 1522 or 1523, he went to Solothurn, Switzerland, to team up with Farel. The exiled Duke Ulrich appointed Gayling as his court preacher in Mömpelgard (Montbéliard). In all probability the knight Dietrich of Gemmingen arranged this alliance.[2]

Among the few communities in northern Baden that had a residing clergyman who changed religious affiliation were Bruchsal and Durlach. They need to be examined more closely. In Bruchsal, Anton Engelbrecht and Johann Stumpf became converts in 1520 and 1522 respectively. Stumpf moved to Switzerland shortly after he joined the new church. As son of a tanner from the Odenwald, he moved to Strasbourg and Heidelberg in 1515 and 1517, and enrolled at the university in 1519. It was here that he had the first real contact with Luther's ideas; it is almost certain that he took part in the 1518 disputation, and visited Luther at his quarters afterward. Among the younger teachers and fellow-students he counted Johannes Brenz, Theobald Billican, and Christoph Siegel as his friends. Brenz and Billican openly confessed Luther's faith in their lectures after they were spelled out in the Augustine convent.

In February, 1520, Stumpf entered a position in the episcopal consistory in Speyer, where the suffragan and parson at the Bruchsal parish church, Anton Engelbrecht, further educated him in the Lutheran theology. One year later he joined the Order of St. John and attended its college in Freiburg/Breisgau. Dissatisfied with his studies there, he sought close contacts with Anton's brother Philipp Engelbrecht the poet and classics professor at the university.[3] It will

---

[1] Otto Conrad, "Johann Gayling von Ilsfeld, ein Reformator Württembergs," Blätter für Württembergische Kirchengeschichte, N.F. XLIII (1939), 13-27.

[2] Ibid., p. 19.

[3] Ernst Gagliardi, et al., Johannes Stumpfs Schweizer und Reformationschronik (Basel: Birkhäuser, 1952), I, vii-ix.

be recalled that he was by far the most important opinion leader for the new group in Freiburg. After Stumpf's ordination in Basle he was transferred to Bubikon near Zurich. There he actively taught the new gospel.[1]

In the meantime, the Speyer suffragan and clergyman Anton Engelbrecht sought to reform the religious situation in Bruchsal. Like his older brother in Freiburg, he had attended Wittenberg university (1515?) and earned his Master of Arts degree in Basle. Thereupon he was appointed chaplain in the cathedral of Speyer. In June 1519, he received a doctor of theology.[2] A Lutheran since 1520, Anton Engelbrecht began to expound the Lutheran doctrine in Bruchsal's parish church in 1522. Two years later, the Speyer cathedral chapter tried and dismissed him. In virtue of the large number of followers he had won in the town and country however, the magistracy and population of Bruchsal vigorously protested the Speyer ruling.[3] As soon as he found a preaching position in Strasbourg, he managed to convert his chaplain, who also lost his position. Even Engelbrecht's successor, Anastasius Meier, formerly a chaplain of the ecclesiastical court, adopted the Lutheran faith.[4]

Southern Franconia

The Heilbronn reformer, Johann Lachmann, was closely associated with the university of Heidelberg. While there, he established contact with a group of theologians who also became reformers in northern Württemberg in the early 1520's. Lachmann's role in Heilbronn's reformatory movement is discussed later in the chapter devoted to preaching foundations.

In Obergriesheim, located between Wimpfen and Mosbach, a parson by the name of Martin was accused by the authorities of assisting the rebelling peasants in their assault against the local headquarters of the German Teutonic Knights at Gundelsheim, and also of being a follower of Luther. He denied the

---

[1] Ibid., p. ix.

[2] Wackernagel, Die Matrikel der Universität Basel, p. 334; and Wilhelm Vischer, Geschichte der Universität Basel von der Gründung 1460 bis zur Reformation 1529 (Basel: Georg), p. 230.

[3] Krebs, Die Protokolle des Speyerer Domkapitels, Vol. II, No. 6256, pp. 1518-31.

[4] Franz Remling, Geschichte der Bischöfe zu Speyer, II (Mainz: Kirchheim), 251; and Kattermann, Die Kirchenpolitik Markgraf Philipp I, p. 71. Meier left for Steinbach after Philipp embarked on a more conservative ecclesiastical policy, with virulent attacks against Meier's innovation in the sacramental system.

first charge but admitted the second.[1] Nevertheless, had it not been for the Peasants' War, he probably would have missed his rendez-vous with Lutheranism.

Parson Johann Herolt of Reinsberg near Schwäbisch-Hall became the first Lutheran preacher in the territory of that imperial city. He had visited the Latin School in Schwäbisch-Hall, then studied in Tübingen, where he received his M.A. degree. His mutual contact with Johannes Brenz, together with his reading of Luther's works made him a convert and a signer to the Syngramma suevicum, the joint declaration of the Lutheran preachers in Franconia. This document was one of the most important items in the Zwinglian-Lutheran communion dispute of October 21, 1525. In spring of that year, Herolt had abolished the requiem and afterward the mass, substituting a second sermon on Sunday afternoon for both. Besides contact with Brenz, the pressure exerted by dissatisfied, rebelling peasants was enough to produce a change of mind in Herolt. On April 2, 1525, the peasants demanded a preacher who stood for evangelical freedom. Their local leader called on Herolt to "establish the evangelism."[2]

In Orendelsall, north of Schwäbisch-Hall, Wolfgang Taurus is said to have been involved with Luther's theology and to have been converted in 1517. But this date seems much too early. Taurus enrolled at the University of Heidelberg and received a master's degree in 1511. Sigel's suggestion that he actively preached the new gospel between 1518 and 1524 is difficult to accept,[3] since the ruler of Hohenlohe, who opposed the Reformation, took no steps to curb such heresy. It can be conjectured, however, that by signing the Syngramma suevicum on October 21, 1525, Taurus was already an active convert by the time the Peasants' War was in full swing. And he was still at it as late as 1532, when he attended another meeting in Heilbronn concerning the Lutheran communion.[4] It was almost inevitable that the Orendelsall preacher, together with

---

[1] Beschreibung des Oberamts Neckarsulm (Stuttgart: Kohlhammer, 1881), p. 211.

[2] Gustav Bossert, "Johann Herolt," Württembergische Vierteljahrschrift für Landesgeschichte, IV (1881), 289-95; and Gerd Wunder, "Georg Widman (1486-1560) and Johann Herolt (1490-1562): Pfarrer und Chronisten," in Lebensbilder aus Schwaben und Franken (Stuttgart: Kohlhammer, 1960), VII, 41-51.

[3] Sigel, Das evangelische Württemberg, Vol. VI, No. 1, p. 231.

[4] Bossert, "Die Syngrammatisten," p. 21.

neighboring Ingelfingen's should have been influenced by Johannes Brenz in Schwäbisch-Hall in the same year, since Brenz had come into prominence among Lutheran intellectuals as early as 1523 in places as far away as Ulm.[1] And it was Brenz who helped Taurus in the early 1530's to his new position in Gross-Sachsenheim near Bietigheim.[2]

In Tauberbischofsheim, Christoph Schreiber[3] was the only clergyman who sided with Luther soon after he burned the Bull <u>Exsurge domine</u>. Schreiber persuaded his chaplain to break with the traditional church shortly afterwards. But he did not command the solid support of the local burghers as some of the other preachers did, for example Engelbrecht in Bruchsal, and Alber in Reutlingen. In retaliation, Schreiber was promptly removed from office in early 1522 by the Mainz archbishopric and the local magistracy.[4]

A backlash against this counter-reformatory activity occurred in 1525. The burghers joined the rebelling peasants in capturing Boxberg and Schweigern; but with their defeat at Königshofen on June 3, the chances for realizing Luther's ideas were as far removed as before.[5]

It is difficult to assess who influenced Schreiber. Where he received his B.A. or M.A. is not known. Although he may already have been influenced by his early student colleagues, it is almost certain that he had contact with the Lutheran clergyman in Miltenberg, Johannes Drach.[6] After fleeing that city, Drach addressed a letter to his followers, asking them to pray "for our brothers Antonium (Chaplain Scherpfer in Miltenberg) and Christophelen, parson at

---

[1] Gustav Bossert, "Zur Brenzbiographie," <u>Blätter für Württembergische Kirchengeschichte</u>, X (1906), 107, footn.

[2] Kurt Bachteler, <u>Geschichte der Stadt Gross-Sachsenheim</u> (Bietigheim: Gläser & Kümmerle, 1962), p. 83.

[3] Listed as Cristoferus Scriptoris from Lauda in Toepke, <u>Die Matrikel der Universität Heidelberg 1386-1662</u>, I, 496. He enrolled at Heidelberg university in July, 1514.

[4] Wilhelm Ogiermann, "Tauberbischofsheim im Mittelalter; Urkundenforschung zu Kultur und Geschichte im Zeitraum von 800-1600," <u>Tauberbischofsheim: Aus der Geschichte einer alten Amtstadt</u> (Tauberbischofsheim: Fränkische Nachrichten, 1955), p. 289.

[5] Erich Keyser, ed., <u>Badisches Städtebuch</u> (Stuttgart: Kohlhammer, 1959), p. 157.

[6] Drach was a cousin to Friedrich Weigand, the revenue official (<u>Amtskellar</u>) of the Mainz electorate in Tauberbischofsheim.

Bischofsheim . . ."[1]

In Rothenburg/Tauber and Würzburg, two average-sized cities, Lutheran principles and views were spread by the occupants of the preaching foundations. How and where this change originated is discussed in Chapter VII.

South Baden

The university town of Freiburg/Breisgau, with a population of close to 6,500, commanded a powerful and determined Austrian-Roman Catholic party that did everything in its power to prevent a cleavage from the established church. Nonetheless, Lutheranism gained some following among teachers and students. The great majority of Freiburgers, however, remained aloof, even though they had just experienced a bad harvest with price increases and the sale of indulgences in 1517 and 1518. The social distance between the academic community and the general population was an additional reason for the limited curiosity. This enabled the city government to be far harsher against the Lutherans than against the university senate. It even sent representatives to the emperor to accuse certain academic persons of being renegades. Still, the prevailing direction in the community was toward the new faith,[2] so that a number of Freiburgers asked their bishop in Constance early in 1522 for permission to practice communion in both forms, which he rejected. Hence, all Reformation matters emanating from Freiburg came strictly from the university, whose effectiveness was greater in more distant regions of the Southwest than it was in the city or the neighboring hinterland. The impact was especially felt in the southern tier of Upper Germany, including the Austrian territories that lay scattered from Tyrol to Alsace. Moreover, Freiburg's was the only university whose graduates avowed the Zwinglian branch of Protestantism after 1525.

The humanist and poet, Philipp Engelbrecht was a vigorous advocate of the Lutheran dogma for over half a decade. Since his graduation from Wittenberg he had known Luther as a teacher, and being the only person in Freiburg to supply first-hand information about the prestigious reformer, many people sought his opinion and advice.[3] In a memorandum placed on the bulletin board

---

[1] Vierordt, Geschichte der Reformation im Grossherzogtum Baden, p. 141.

[2] This must have been left largely to the natives of the city such as Thomas Sporer, who harbored Lutherans and disseminated propaganda.

[3] Albert, "Die reformatorische Bewegung zu Freiburg bis zum Jahre 1525," p. 6.

in 1521, Engelbrecht urged students to buy Luther's works; and in lectures on Virgil, he praised the Wittenberg reformer as the most oustanding apostle of Christ.[1]

Engelbrecht cultivated close ties with Urbanus Rhegius in Augsburg, who, at Christmas 1520, came to Zähringen near Freiburg and remained for six months. There he and his Protestant friends frequently met in Engelbrecht's house. We have no information on what went on in these meetings, although we can surmise on the basis of subsequent events that they reinforced each other's religious convictions. On September 5, 1521, the academic senate of the university confronted Engelbrecht with his pro-Lutheran sentiments and warned him to immediately stop such "tricks" (Possen).[2]

Among Ulrich Zasius' associates who shared the same conviction were Bonifacius Amerbach, Johannes Zwick, Thomas Blarer, Hieronymus Vehus, Wolfgang Capito, Jakob Otter, and Caspar Hedio. Zasius, a professor of law, initially supported Luther's canons with enthusiasm, but rejected them as soon as they gained hold among the masses in 1523-1524. Zasius' younger friends left Freiburg long before there was any noticeable support by men like Engelbrecht, Otter, Zell, and Öler for the Lutheran creed. The fact that all departed when the political authorities clamped down on the new teaching proves that they took their faith seriously.[3]

The town of Lahr had a preachership (Prädikatur) that was established at the end of the fifteenth century. Its occupant, Jakob Ofner, and parson Martinus N.[4] were dismissed in 1528.

In Hecklingen, south of Kenzingen, Johann Kress (also called "Parson Hans") seems to have been affected by the strong reformatory spirit exemplified by Johann Otter in the neighboring town. Following the forceful suppression of the movement there, he gave up his office and sought refuge in Stras-

---

[1] Hagenmaier, Das Verhältnis der Universität Freiburg i. Br. zur Reformation, p. 15.

[2] Ibid.

[3] Matthias Zell, who likewise received his education at the University of Freiburg, was chaplain in the cathedral and head of the Pfauenburse (Prep. School), and finally rector of the university until 1517-1518. After his term in office, he became pastor and foundation preacher at the cathedral in Strasbourg. He kept his benefice in Freiburg for a few more years, probably until 1524.

[4] In the sixteenth century, the initial N. was equivalent to X, thereby pointing to the absence of a family name.

bourg.[1]

The villages of Wittnau and Ballrechten, south of Freiburg, likewise maintained Lutheran preachers. At what time they began to indoctrinate their community is difficult to assess; the chaotic period of the Peasants' War seemed to have offered the opportunity to get rid of the hierarchical Roman-Catholic superstructure. In Ballrechten, parson Jakob Stümpfli fled in 1525 for fear of being punished for the open commitment that he displayed to Luther in his sermons.[2] In neighboring Wittnau, Johann Wehe, a former friar of St. Ulrich's and parson at Bollschweil, switched sides during the Peasants' War. However, the identity with the preacher and peasant leader in Leipheim, the cousin of Eberlin of Günzburg, can not be determined.[3]

The dean of the Breisach rural chapter and parson of Schlatt, Peter Spengler, openly complained about conditions in the Catholic Church. He even read Lutheran books and communicated with Lutheran ministers. One night, the Catholics raided his apartment, arrested him, and accused him of being secretly married as well as trying to convert the parsons in his chapter.[4]

Summary

From this discussion a number of conclusions can be offered. (1) The interior regions of southwestern Germany, no matter whether of territorial or imperial status, were influenced from the border areas; they were linked to each other by academic and business ties. (2) The upper Rhine cities of Freiburg and Strasbourg had a strong influence on the upper and middle Neckar cities. Esslingen received additional assistance from Ulm and, in turn, became an important diffusion center for towns and villages in the surrounding Duchy of Württemberg. (3) The city of Constance, located centrally at the southern end of our area, was ideally suited as a natural harbor for the new religion and its followers and its influence radiated out to the other cities in Upper Swabia. (4) It seems clear that no preacher proclaimed the new faith

---

[1]Albert, "Die reformatorische Bewegung zu Freiburg bis zum Jahre 1525," p. 19.

[2]F. Brunner, "Historische Notizen über die Pfarrei Ballrechten," Freiburger Diözesan-Archiv, XIV (1881), 289.

[3]Albert, "Die reformatorische Bewegung zu Freiburg bis zum Jahre 1525," p. 74.

[4]Vierordt, Geschichte der Reformation im Grossherzogtum Baden, pp. 279-80.

unless there was a considerable demand by the people. It was a true "movement" that began in the community and, with time, reached its governing body.

Furthermore, it can be concluded that resident preachers who embraced the new faith were not as devoted to the new cause as immigrant or itinerant preachers. It is also evident that a direct personal influence stemmed from Zwingli in Zurich toward the reformers of Stein, Waldshut, and Schaffhausen. Finally, personal ties and institutional connections that antedate Luther's 95 Theses played an important role in the subsequent spread of his ideas.

CHAPTER VI

LOCALITIES WITH IMMIGRANT PREACHERS

In the preceding chapter, attention was directed to the various localities with resident preachers who switched to the new faith while performing their duties. The study now turns to an examination of immigrant preachers who moved to different localities after embracing the Protestant faith. Their contribution to the diffusion of Lutheranism is difficult to over-estimate.

In general, immigrant preachers were associated with smaller towns and villages which, in turn, depended on larger towns and cities for such personnel. An immigrant preacher could be either a convert from the neighboring city, or someone who had been preaching in the city. He also could have come from a more distant place and have been channeled into a village via a neighboring city. In any case, city and village churches were connected via the patronage system through which the magistracy, monastery, or hospital (Spital) had the presentation right (ius praesentandi) over the village church. In cases where the magistracy ordered the Reformation of the urban monasteries or hospitals, and for these, in turn, to reform their institutions outside the city, one can speak of a wave-like migration created by the neighborhood effect.

Preachers who were native to their locality and had just returned from another city, where they had adopted the new ideology, were often in a more influential position than those preachers who were new. Where the predominant city or territorial ruler, after a short evangelical phase, turned strongly anti-Protestant (such as in Freiburg, Rottweil, Heidelberg, and Schwäbisch-Gmünd), the direct or indirect impact of the movement on the umland and hinterland were negligible.

Itinerant Preachers

A special group engaged in converting the illiterate populace were the itinerant preachers. They carried the evangelical message from one locality to another. Whereas resident preachers left an indelible imprint on their congregations, itinerant preachers preached in a place for short periods only. Hence their duty and purpose was to make the population familiar with the new ideas, rather than to organize and head a new group. They preached mostly to a limited audience in private homes or under the open sky in cemeteries, fields, and outlying monasteries on quickly assembled pulpits provided by the peasants or townsfolk. Since these preachers often represented the very beginning of the Reformation movement, they were of major significance in getting it off the ground.[1] Some of them enjoyed such popularity that they were preferred over the succeeding regular preacher chosen by the city council.

However, the Austrian rule in Württemberg, Breisgau, Hohenberg, and Burgau made it impossible for most migrant preachers to stay at one place for any extended period of time. As early as 1519, the Bavarian dukes issued an order outlawing them, together with street robbers and gypsies. While many were uneducated, not to say boorish, a number of capable, educated migrant ministers well versed in the Scripture appeared under the cloak of craftsmen or peasants. Often they carried the common title "Karsthans," which enabled them to establish close contact with the man in the street. Such men as Johann Murer, Eberlin of Günzburg, Johann Locher, and Wolfgang Vogler often traversed long distances between their place of conversion and their later assignment. Information on itinerant preachers is exceedingly scarce during this period because comparatively little appears in the permanent records of towns or local churches. However, sufficient evidence is available to permit the following case studies.

Doctor Johann Murer, dressed in peasant garment and nick-named "Karsthans," preached in the upper Rhine and Neckar reaches. As a graduate of the University of Freiburg, he moved around in neighboring Breisgau and Alsace before preaching seriously in his native Horb as well as Haiterbach between December, 1522 and March, 1523. Thereafter he continued his mission in Balingen, but was soon jailed by the Austrian government.[2] Support for the movement also came from Sebastian Lotzer, a well-known furrier and peasant leader in Memmingen, who encouraged the citizens in his native Horb to reject the old church. In May, 1524, a young organist and prebendary, Konrad Sterzler, extolled the virtues of the evangelical creed through his sermons.[3] In the Baar

---

[1] Albert Angele, ed., _Altbiberach um die Jahre der Reformation_ (Biberach: Renner, 1962), p. 190.

[2] Paul Kalkoff, "Die Prädikanten Rot-Locher, Eberlin und Kettenbach," _Archiv für Reformationsgeschichte_, XXV (1928), 131.

[3] Gustav Bossert, _Aus Horb a. N. und Umgebung_ (Horb: Christian, 1936), p. 127.

region immediately to the south, a thoroughfare between Württemberg and Switzerland, Peter Bach was jailed in 1530 for preaching "revolutionary" ideas.[1]

After his expulsion from Ulm, Eberlin of Günzburg went to northern Switzerland to administer the new faith to the people of Baden. He then went to Wittenberg for one year, returning to Basle and Rheinfelden in summer 1523, to submit several of his works for publication. On request he began to preach there, soon generating a lively interest and participation from magistracy, citizens, and peasants. Again, this was to bring about his expulsion by the Austrian government. He then went to Rottenburg and Brackenheim, where he met with the local preacher, Konrad Sam. The Austrian ruling class clearly sensed danger in their meeting and told them to leave Brackenheim. After a short stay with the Knight of Sickingen, Eberlin returned to Saxony and, on Luther's recommendation, came to Wertheim/Main to serve as reformer to Duke Georg II.[2]

In Bopfingen, near Nördlingen, Wolfgang Vogel, an itinerant preacher from Reutlingen and graduate of the University of Freiburg[3] soon was asked to chair the growing Protestant party after delivering some sermons to crowds in the square. Due to the spillover effects of Billican's sermons in Nördlingen, the Bopfingen residents soon demanded the permanent hiring of Vogel for delivering sermons in the German language. But the local town council, aware of the emperor's hostility toward everything Lutheran, rejected the bid and waited for further developments. Vogel shortly afterward left for Eltersdorf, near Nuremberg. Jakob Yedler, elected by the city six months later, continued in Vogel's position; yet he proved to be not nearly as popular as his predecessor.[4]

Two others, Johann Locher and Heinrich Kettenbach, experienced the authority of their mutual friend Eberlin von Günzburg in the Ulm monastery. Locher left that institution in 1524. Complete clarity of his religious conviction

---

[1] Lauer, "Die Glaubenserneuerung in der Baar," Freiburger Diözesan-Archiv, XIX (1919), 95.

[2] Born 1465 in Günzburg near Ulm, he went to the universities of Basle and Freiburg, where he earned his M.A., and then entered the Franciscan order. In 1519 he was still preaching in Tübingen in the old tradition. Shortly after he came to Ulm, he read Luther's writings, leading to his break with the old church. Max Radlkofer, Johann Eberlin von Günzburg und sein Vetter Hans Jakob Wehe von Leipheim (Nördlingen, 1887).

[3] Mayer, Die Matrikel der Universität Freiburg im Breisgau von 1460-1656, p. 215.

[4] Friedrich Richter, Zwei Schilderungen aus der Geschichte der ehemaligen Reichstadt Bopfingen (Nördlingen: Beck'sche, 1862), pp. 2-3.

and realization of his duty to spread the new gospel came after a long and tedious itinerary. He was also a member of a circle of significant men whose faith was already well entrenched. He soon left Ulm for Augsburg, where he broke the monastic vow upon observing how rapidly the Franciscan monastery there was dissolving.[1] Locher spent most of his time in Zwickau, but was also active in spreading the new theology between Lindau and Augsburg, Hesse and Donauwörth. His frequent reference to <u>Karsthans</u> in his Zwickau letters does not necessarily apply to Johann Murer, since many popular preachers carried that epithet.[2]

One last example, Diepold Peringer,[3] an expelled Benedictine monk from a cloister near Gundelfingen, proselytized with great force in the villages of Thon and Wöhrd, near Nuremberg, as well as in the city itself. When the magistracy forbade him to hold his popular sermons, he departed for Kitzingen and Rothenburg/Tauber. It is noteworthy that a great many of the itinerant preachers either moved to or from Ulm.

The Ulm Territory

The environs of the imperial city of Ulm possessed a number of localities that depended on a preacher coming from the outside. For example, in Geislingen, Paul Beck appeared in late 1526 to conduct evangelical sermons after he had been expelled by the Swabian Federation for this activity in his native Munderkingen. He had been a student in Heidelberg[4] when Luther held his debate in front of the Augustinian order. But how did Beck come to his position in Geislingen? The first evidence of a growing Protestant movement there was on April 20, 1526, when the Ulm magistracy sent a complaint and warning to Geislingen's Catholic town parson, Dr. Georg Oswald, for labelling the preacher in

---

[1] Paul Kalkoff, "Die Prädikanten Rot-Locher, Eberlin und Kettenbach," p. 38.

[2] Ibid., p. 145.

[3] Enrolled in Wittenberg as Johann Bermiger from Kitzingen on March 19, 1520. See Carolus E. Foerstemann, <u>Album Academiae Vitebergensis 1502-1560</u> (Leipzig: Tauchnitz, 1811), p. 88.

[4] Beck enrolled on January 13, 1516, and earned a B.A. (via moderna) in July, 1523. After his discharge in Munderkingen in March, 1526, he returned for one year to Heidelberg as chaplain and assistant before being appointed as a preacher in Geislingen.

Ulm, Konrad Sam, a "heretic" and life in Ulm as "Turkish, bestial and devilish."[1] Oswald was also accused of withholding sacraments to a number of devoted Geislingen citizens because they possessed or had read Luther's translation of the New Testament.[2]

On December 7, 1526, a group of 39 Geislingen burghers petitioned the Ulm magistracy to provide them with an evangelical preacher. Among the signers were the long-established and respected families Mördlin, Widmann, Schollkopf, Hennenberg, Sattler, Bürgermeister, and von Naw. This was further indication that the Reformation was espoused first by groups that enjoyed a socio-economic advantage. Although the 39 signers were a minority, the city council in Ulm complied with their demand to have Paul Beck as their preacher.[3] He delivered his first sermon in Geislingen on April 27, 1527.

It has been said that Beck became preacher in Geislingen through the agency of Konrad Sam, who at the time was the foundation preacher in Ulm.[4] Although we lack details regarding this affair, there may be some truth in the claim.[5] Neither do we have any information on whether Wolfgang Rychard, who was born in Geislingen and served as doctor and humanist in Ulm, was vital in establishing rapport between Beck and the Protestants in Geislingen. In any case, Rychard played an important role in spreading reformatory ideas in and around Ulm. He had intensive contacts with its monasteries, of which he became the secret supplier of reformatory literature and cartoons.[6]

Despite the resolute opposition of the Catholic faction led by Dr. Georg Oswald, Geislingen became a secondary diffusion center for the western section of Ulm's imperial territory, also called Helfenstein. For example, seven men from adjacent Kuchen were among the 39 signers of the Geislingen petition mentioned above. Among these were the bailiff, an innkeeper, and a customs offi-

---

[1] Georg Burkhardt, Geschichte der Stadt Geislingen an der Steige (Konstanz: Thorbecke, 1963), pp. 179-80.

[2] Ibid., p. 179.   [3] Ibid., p. 180.

[4] R. Burkhardt, "Paulus Beck: Der erste evangelische Geistliche Geislingens," Blätter für Württembergische Kirchengeschichte, XXXV (1931), 254.

[5] Konrad Sam was born in Rottenacker, located only $2\frac{1}{2}$ miles further down the Danube from Munderkingen, the birthplace of Paul Beck. Sam was the older and possessed an advanced degree.

[6] Friedrich Keidel, "Johannes Piskatorius," Blätter für Württembergische Kirchengeschichte, VI (1902), 143-78.

cer,[1] all persons of high socio-economic status or associated with the transit traffic on one of the busiest thoroughfares in all of Germany. The local Helfenstein customs house was an extra source of income for the community. It collected not only the same amount of customs duties as nearby Geislingen, but also profited from offering lodgings to carriers who stayed overnight.[2]

Nevertheless, only the systematic introduction of the new faith by Ulm in 1531 brought a new parson to Kuchen, namely Simon Rait of Lauingen. The Catholic parson was unwilling to change in response to the demands of the Zwinglians in his parish. Consequently he was discharged and went to Schwäbisch Gmünd. The bailiff and citizens of Kuchen were well satisfied with Rait's performance.[3] Even the inhabitants and visitors of neighboring Überkingen, known for its spa, attended Beck's sermons some years prior to their official initiation, despite the two-hour trip to and from Geislingen.[4]

In Pfuhl, east of Ulm and across the Danube, the Ulm magistracy allowed a group of eminent[5] persons to hire their own preacher. His salary was to be paid out of their own pockets.[6] Yet, unlike the situation in Geislingen, Leipheim, Langenau, and Reutti, we have no knowledge of a hired clergyman prior to the second half of 1531. However, Pfuhl's parson, Philipp Neidlinger claimed to have preached evangelism to the people of Langenau before the large scale missionary efforts referred to above. The local preaching foundation,

---

[1]Georg Burkhardt, "Evangelisches Leben in Geislingen vor der Reformation," Geschichtliche Mitteilungen von Geislingen und seiner Umgebung, IV (1933), 10.

[2]Beschreibung des Oberamts Geislingen (Stuttgart: Cotta'schen, 1842), p. 216.

[3]Friedrich Keidel, "Ulmische Reformationsakten von 1531 und 1532," Württembergisches Vierteljahresheft für Landesgeschichte, N. F. IV (1895), 279; and Sigel, Vol. IV, No. 2, p. 848.

[4]The local mineral water for drinking and bathing was preferred by the citizens of Ulm, and visitors from Upper Swabia, especially the wealthier sections of Augsburg, Dillingen, Memmingen, and Switzerland. See Beschreibung des Oberamts Geislingen, p. 245. In the sixteenth century, Überkingen was the favorite spa especially of the nobility, many of them Catholics.

[5]In the protocol, the adjective sondern may be interpreted as referring either to an isolated or especially important, eminent group of persons, or both. See Hermann Fischer, Schwäbisches Wörterbuch, Vol. V, column 1445.

[6]R. Burkhardt, "Paulus Beck: Der erste evangelische Geistliche Geislingens," p. 250.

founded in 1468 by Peter Arnold, a priest and citizen of Ulm, was of limited or no importance in introducing reformatory points of view. Therefore, the success of the movement was intimately tied to the missionary activities of Ambrosius Blarer, Oecolompadius, and Bucer from Constance, Basle, and Strasbourg, respectively.

In Reutti, Johann Mann (Mang) declared that he had been to some extent "illuminated" since 1529, very likely through Beck in adjacent Geislingen. He pledged his wholehearted support to Ulm's 18 Reformation Articles. Since the abbot of Blaubeuren demanded that he abide by the traditional ceremonies, Mann asked the magistracy of Ulm to appoint him as the evangelical preacher in Stubersheim.[1] On May 30, 1532, Jakob Muntprat (Mundbrot) came from Constance to Reutti to take over the preaching post.[2]

On September 24, 1532, the old parson[3] of Türkheim, who had been condemned by the reformers as "a poor, miserable Popist," was replaced by the Protestant Simon Vogler (Barter). He quit on September 24, 1532, because his people were upset about him and the new faith. As he wished to resume the occupation he had previously held,[4] he was replaced by a preacher from Riedlingen.[5]

Neighboring Aufhausen was served by the preachers from Türkheim and Deckingen (the territory of Weisensteig), one Protestant, the other Catholic. Judging by the attendance records, everybody went to hear the former's sermons and boycotted the Catholic service.[6]

In Böhringen, north of Überkingen, the inhabitants expressed their gratitude to the Ulm magistracy for their new preacher, Martin Karter.[7] Despite being subjected to severe punishment, he received many visitors from neighboring Helfenstein and Württemberg.[8]

---

[1] Sigel, Das evangelische Württemberg, Vol. VI, No. 2, pp. 658-59.

[2] A Jakob and Johann Muntbrot studied 1528 in Tübingen; Heinrich Hermelink, Die Matrikel der Universität Tübingen (Stuttgart: Kohlhammer, 1931), I, 262.

[3] Mathias Schreiber, enrolled at the University of Freiburg in 1506.

[4] Keidel, "Ulmische Reformationsakten von 1531 und 1532," p. 339.

[5] Sigel, Das evangelische Württemberg, Vol. VII, No. 2, pp. 1025-27.

[6] Keidel, "Ulmische Reformationsakten von 1531 und 1532," p. 319.

[7] Ibid., p. 298.

[8] Ibid., p. 317.

Interior Württemberg

In Reichenbach, east of Esslingen, parson Peter Rieker actively participated in the Peasants' War.[1] For unknown reasons he was permitted to practice his occupation after their defeat. He had enrolled at Tübingen university in January, 1518,[2] and became parson in Reichenbach in early 1525. What Rieker did between these dates, except studying, is unknown, but the assumption that he had close contact to the Reformation sources in his hometown Esslingen is more than mere speculation. Even before the peasant uprising he had been an opponent of the mass and the small tithe. Following the peasants' defeat, he openly predicted that, according to the Bible, another revolt would take place in 1527, resulting in the final coming of evangelism. Anabaptists in Esslingen and Augsburg voiced similar prophecies.[3] When the Austrian government expelled him for his anti-Catholic stand, he took up a preachership in Sexau, north of Freiburg.[4]

In Schwäbisch Gmünd, Johann Schilling, a Barefoot monk from Rothenburg/Tauber began to communicate the new teaching in 1523 with a strong social-revolutionary bent. He was soon driven out by the magistracy and moved to Augsburg.[5] He was succeeded by Andreas Althamer, who may have turned Protestant while studying in Leipzig from 1519 to 1521. It was there that he witnessed the sensational disputation between Luther and Eck. After serving one year at the Latin School in Schwäbisch-Hall, he returned to Reutlingen in spring 1522. From there he went to Gmünd, where he became chaplain.[6] Notwithstanding prohibition orders by his superior, he began to disseminate the new ideas, leading to a split in the parish, with one faction favoring Althamer, the

---

[1] Wilhelm Böhringer, ed., Heimatbuch Reichenbach an der Fils (Reichenbach: Schaul & Schniepp, 1968), p. 64. Rieker joined Matern Reuernbacher's peasant army as it marched towards nearby Kirchheim-Teck.

[2] Hermelink, p. 219, has him listed as Retrus Rieder from Esslingen.

[3] Gustav Bossert, "Beiträge zur Reformationsgeschichte Württembergs," Blätter für Württembergische Kirchengeschichte, XI (1907), 98.

[4] Neu, Pfarrerbuch der evangelischen Kirche, II, 487.

[5] Joseph Zell, "Andreas Althammer als Altertumsforscher," in Württembergisches Vierteljahresheft für Landesgeschichte, N.F. XIX (1910), 429-44.

[6] Emil Wanger, "Die Reichsstadt Schwäbisch Gmünd in den Jahren 1523-1525," Württembergisches Vierteljahresheft für Landesgeschichte, N.F. II (1879), 27-28.

other the newly appointed city parson, Ulrich Schleicher, a Catholic. With the rebuff of the peasants, the Lutheran community asked Nuremberg, Nördlingen,[1] and Dinkelsbühl for advice on how to formulate a new church order. By marrying a girl from Schwäbisch Gmünd in June of 1525, Althamer broke with the old faith once and for all. Not long thereafter, Ludwig Sigwin from Ravensburg appeared in Gmünd and attempted to spread the Zwinglian Reformation. But he deplored the popular attitude about him that he was a Schwärmer,[2] an indication that Gmünd's Lutheran foundation must have been firmly rooted.

Six miles down the Rems, in Lorch, Jeremias Mayer was suspected of being an apostle of Luther's faith in 1524. It seems quite certain that the popularity of Schilling and Althamer had an influence on Mayer, but a more crucial factor was his education at Freiburg university, where he met Philipp Engelbrecht, Thomas Blarer, and Caspar Hedio, all matriculating in 1514.[3] Mayer was threatened on suspicion of being a Lutheran, but through the intercession of friends and members of the Imperial Court in Esslingen, he escaped punishment.[4]

The Ries

The imperial town of Nördlingen, including the villages and towns surrounding it, became Protestant through the efforts of missionary preachers. When Nördlingen's representative to the Worms diet returned in 1521, citizens there as well as in neighboring Bopfingen requested the installation of a Lutheran church service in their towns. For example, Caspar Kantz, a native of Nördlingen, was lecturer with the Carmelite monks in Augsburg. Their cloister served as quarter for Luther's fourteen-day stay in the city in October, 1518. The prior at the time was Johann Frosch, who knew Luther from Wittenberg and welcomed his beliefs. He influenced Kantz to prolong his abode for another

---

[1] Billican advised Althamer to calm the rebellious city groups (in seditionis civicas potissimum) through sermons and publications. See Theodor Kolde, "Zur Geschichte Billicans und Althamers und der Nördlinger Kirchenordnung vom Jahre 1525," Beiträge zur Bayerischen Kirchengeschichte, X (1904), 29.

[2] Emil Wagner, "Die Reichsstadt Schwäbisch Gmünd in den Jahren 1526-1530," Württembergisches Vierteljahresheft für Landesgeschichte, N. F. IV (1881), 82-83.

[3] G. Hoffmann, "Reformation und Gegenreformation im Bezirk Welzheim," Blätter für Württembergische Kirchengeschichte, XIV (1910), 31-32.

[4] Ibid., p. 31.

year before returning to Nördlingen in 1519.[1] Once returned, Kantz set about proclaiming and defending Lutheran doctrine, first in secret, closed meetings, then in open sermons. His impression on the two largest and most influential guilds in the city, the loden cloth weavers and the producers of fine cloth (Feintuchmacher) was so strong that they, too, became solicitors of religious change. Kantz had special rapport with the loden cloth makers, since his father was a member of their guild before giving up his trade to become a miner in the Ore Mountains. In 1522, Caspar Kantz published his sensational evangelical mass. It was read in German, and it was the first such celebration in Germany. In the same year, Theobald Billican was hired from Weil der Stadt to occupy the preaching foundation that had been established at the parish church St. George.[2]

Nördlingen was in many respects the "opinion leader" and model for the small neighboring towns of Bopfingen and Harburg. This observation pertains especially to ecclesiastical-reformatory issues raised by Kantz and Monninger. The citizens of Bopfingen voiced their desire for Lutheran church worship, and following the installation of the German mass by Kantz in Nördlingen, requested the same for their town. People envied the large neighboring city for its special preaching foundation, which allowed the reading of sermons by an occupant well-versed in the Scripture. Prior to the recruitment of a standing preacher, Bopfingen's citizens gleaned the Word from a travelling preacher, Wolfgang Vogel, in 1522 and 1524. He was met with such approval by local Lutherans that he gladly gave in to their request to become their permanent pastor.[3]

Vogel's enrollment at the university in Freiburg fourteen weeks prior to Engelbrecht, Blarer, and Hedio leaves no doubt about the latters' significance in converting him to the Lutheran faith.[4] However, the local magistracy preferred to defer Vogel's instatement until Lutheranism had become more popular; it was secretly hoping for moral and political support from the larger cities.

---

[1] Gustav Wulz, "Caspar Kantz," in Lebensbilder aus Bayerisch Schwaben (München: Max Hueber, 1955), IV, 114.

[2] Max Cantz, "Caspar Kantz und die Nördlinger Reformation," Historischer Verein für Nördlingen und Umgebung, Zwölftes Jahrbuch (1929), p. 161.

[3] Richter, Zwei Schilderungen aus der Geschichte der ehemaligen Reichsstadt Bopfingen, p. 2.

[4] A clear indication was Vogel's siding with the "Evangelicals" against the "Lutherans" when the Zwinglian-Lutheran dispute began in Nuremberg and vicinity in 1524. See L. Keller, "Wolfgang Vogel," Allgemeine Deutsche Biographie, XL, 128.

This problem must be seen in light of the sharp opposition by most clergymen, backed by their superior, Bishop Stadion of Augsburg. In spite of heavy pressures, the Protestants persisted in their request for a permanent preacher, offering Vogel as the only choice. The magistracy agreed to the first demand, but insisted on hearing one or two more preachers in order to see whether their sermons would coincide. In summer, 1524, the city government chose Jakob Yedler from Nördlingen over Vogel.

The fact that Yedler was a "native" and therefore entitled to the office of standing preacher rather than Vogel, who was from Reutlingen can hardly be the sole explanation for the magistracy's decision. The latter's insistence on comparing Vogel's ideas with those of other Protestants strongly suggests an already existing division within the Protestant party. Since Vogel represented the Strasbourg-Zurich axis, Yedler was regarded as the champion of the Lutheran cause. This turning of the tables by the magistracy was not limited to Bopfingen but could be observed in Ulm, Reutlingen, Kempten, Memmingen, Wertheim, and Nuremberg as well. In Ulm it was the mayor, Bernhard Besserer, who ordered Konrad Sam to shift to the Zwingli-Bucer side of Protestantism, at the expense of the Lutheran side which Sam had represented since 1520.[1] Vogel's reversion to "left-wing" Protestantism in Eltersdorf, near Nuremberg, supports our earlier hypothesis. But Yedler was given only one year. The town council reacted to the strict orders by Emperor Charles V and threats by Bishop Stadion. It unilaterally cancelled the two-year contract that had already been signed by both sides. Eleven citizens protested the action by exciting an assembly of burghers in front of city hall, but under threat of punishment, they submitted to political pressure.[2] Had the magistracy not intervened, it seems certain that the vast majority of Bopfingen citizens would have wanted to keep Yedler, thus giving their stamp of approval to the religious innovation. Four years later, in 1529, the Protestants were allowed a lukewarm preacher of their own, a Wilhelm Vogler, who later became the city parson. He was not related to the first Bopfingen preacher of the same name.[3]

The villages around Nördlingen were likewise looking to the city for refor-

---

[1] Albrecht Weyermann, "Die Bürger in Ulm der zwinglischen Confession zugetan," Tübinger Zeitschrift für Theologie (1830), pp. 142-53.

[2] Richter, Zwei Schilderungen aus der Geschichte der ehemaligen Reichsstadt Bopfingen, p. 5.

[3] Beschreibung des Oberamts Neresheim (Stuttgart: H. Lindemann, 1832), p. 244.

matory ideas. Nördlingen was the patron for the tiny community of Pflaumloch, located between Nördlingen and Bopfingen. Billican frequently held sermons there and tried to evangelize the villagers, which the rulers of Oettingen vigorously opposed. However, one branch of the Oettingen clan, that of Karl Wolfgang, gradually came to espouse the new faith and appointed Paul Warbeck as the Lutheran court preacher at the Harburg residence. Warbeck, like Caspar Kantz and Martin Monninger, had been a monk in the Carmelite monastery in Nördlingen. Like Kantz, he left there in 1518, but instead of meeting Luther in Augsburg, Warbeck went straight to Wittenberg to study under him. In 1524, Warbeck complied to the request of Earl Karl Wolfgang[1] and left Heidenheim, northeast of Nördlingen, for Harburg.

Thanks to the authority of Warbeck, Harburg Castle became a diffusion center for the earldom of Oettingen. It attracted several reformers from the territory of Ansbach, who followed Warbeck's move. The latter diligently evangelized Harburg and environs, including the local parson and chaplain, Johann Keller and Johann Mendlin. As already mentioned, Nördlingen served as the model for the implementation of Luther's ideas,[2] but political necessity and prudence produced only gradual changes, resulting in some form of "compromise-Catholicism."

Rather than abolish the old church order in a single stroke, the earl and his court preacher aimed at changes within the traditional Roman-Catholic framework. These included more frequent proclamations of the Word, serving communion in both forms, and using the German vernacular at baptism and in singing the psalms.[3] The latter changes were particularly welcome as the people desired a closer, more personal relationship to the ritual.[4] As in neighboring towns and territories, almost no religious innovations followed in the wake of the Peasants' War.

---

[1] He came in touch with the religious movement on several diets he attended, possibly as early as the Worms diet in 1521. In addition, the attendance of one of Osiander's sermons in Nuremberg left a deep impression on him, to be further reinforced by his wife, Elizabeth, whom he married in 1524. See Reinhold Herold, Geschichte der Reformation in der Grafschaft Oettingen 1522-1569, Schriften des Vereins für Reformationsgeschichte No. 75 (1902), p. 3.

[2] Wilhelm Sievers, Über die Abhängigkeit der jetzigen Confessionsverteilung in Südwestdeutschland von den früheren Territorialgrenzen (Göttingen: Peppmüller, 1884), p. 34.

[3] Zoepfl, "Das Bistum Augsburg und seine Bischöfe," II, 52.

[4] Rösler, Geschichte und Strukturen der evangelischen Bewegung, p. 212.

Herolding had a Lutheran minister by 1529, who was chaperoned by Earl Karl Wolfgang. His brother, Ludwig XV, demanded his resignation the same year. However, most villages under the earl's political rule or ecclesiastical patronage were reformed by 1539.[1]

Some twenty miles to the northwest of Harburg were a cluster of Franconian cities, towns, and villages that maintained a Lutheran preacher. The reformers of Schwäbisch-Hall, Crailsheim, Dinkelsbühl, and Feuchtwangen carried out their offices more or less independent of each other until 1525, when Schwäbisch-Hall rose to a dominant position through its foundation preacher Martin Brenz. Schwäbisch-Hall's religious eminence had from the beginning depended strictly on Brenz, as will be discussed in Chapter VII.

Eastern Franconia

The Crailsheim reformer, Adam Weiss, was the son of that town's mayor. He studied at Mainz university, where he subsequently taught from 1512 to 1521 with a licentiate in theology. Weiss, together with other Mainz faculty members, had become an enthusiast of Luther in 1519.[2] He also was a close friend of Caspar Hedio, the highly respected foundation preacher of Mainz. This intimate relationship with the later Strasbourg reformer, combined with his own indomitable humanistic orientation prevented him from endorsing the Syngramma suevicum, which had been signed by Brenz and thirteen other clergymen of northern Württemberg. The admiration which Weiss had for Zwingli and his disciples[3] during the 1523 dispute in Zurich and afterwards, was one more reason why he did not want to belong to Luther's Zürrlimänner[4] and "trivial little bishops" who opposed his choice of communion.[5] In February 1525, Billican admonished Weiss for meddling with the Zwinglian reformers.[6] However,

---

[1] Herold, Geschichte der Reformation in der Grafschaft Oettingen 1522-1569, p. 6.

[2] Steitz, Geschichte der evangelischen Kirche in Hessen und Nassau, I, 49.

[3] Emil Egli, Schweizerische Reformationsgeschichte (Zürich: Zürcher & Furrer, 1910), p. 84.

[4] Also Zürli-Müler, or foolish chatterers. Zwingli's letter to Margrave Philipp I on January 11, 1527.

[5] Gustav Bossert, "Die Syngrammatisten," Blätter für Württembergische Kirchengeschichte, VII (1892), 19.

[6] Julius Gmelin, Hällische Geschichte (Hall: Staib, 1896), p. 733.

Weiss was too geographically isolated to continue functioning in this setting, surrounded as he was by Lutherans.

In November of 1525, Brenz sent his colleague Johann Isenmann to Weiss, carrying a copy of the Syngramma suevicum. Weiss, who had just set up a new church order in Crailsheim[1] considered to be the model by the reformers in Schwäbisch-Hall, was not easily convinced of Brenz' communion concept. Two years later the two men were still corresponding about this matter.[2]

Weiss' sphere of influence did not reach beyond the area immediately west of Crailsheim, which was partly the result of the political and economic superiority of neighboring Hall, with its dynamic reformer. Nevertheless, Crailsheim's view to the east was more favorable, since Weiss was regarded as the outstanding reformer of the thinly populated Franconian corridor between Crailsheim and Ansbach. It was here that Weiss and his colleagues gradually made themselves felt. Eventually Weiss became the personal adviser to the Margraves Casimir and Georg of Brandenburg.

In Altenmünster, the first Protestant sermon was read in 1523, the same year that Crailsheim first heard Weiss. The preacher in Altenmünster was Hans Karpff, who must have had personal contact with Weiss.

Johann Hüfelin of Ellwangen became the preacher of Jagstheim in 1528. A graduate of Tübingen (enrolled on May 4, 1507), he served six years in Jagstheim, until Weiss secured him a better position in Dinkelsbühl.[3]

In Hengstfeld, Balthasar Schnur definitely began to preach the new theology in 1525.[4] He spent most of the three years, from 1518 to 1521, in Würzburg, where he probably belonged to the reformatory group led by Paul Speratus. Schnur gradually and carefully ushered in the new faith against the opposition of the noblemen Wilhelm von Crailsheim and Georg of Wollmershausen.[5]

---

[1] Ibid., p. 738; and Gustav Bossert, "Johann Isenmann," Blätter für Württembergische Kirchengeschichte, V (1901), 147.

[2] Theodor Pressel, Anecdota Brentiana (Tübingen: Heckenhauer, 1868), p. ix. This letter is reprinted in Julius Hartmann and Karl Jäger, Johannes Brentius (Hamburg: Friedrich Perthes), I, 432.

[3] Hermelink, Die Matrikel der Universität Tübingen, p. 159.

[4] Beschreibung des Oberamts Gerabronn (Stuttgart: Cotta'schen, 1847), p. 152.

[5] Gustav Bossert, "Beiträge zur Geschichte der Reformation in Franken," Theologische Studien aus Württemberg, I (1880), 187.

Onolzheim, like Altenmünster, belonged to the Crailsheim parish. The reformatory tide hit this little village in 1525, and in 1530 Johann Gärtner became its first Lutheran minister.[1] Neighboring Rossfeld's fate was to be identical, except that its preacher, Johann Breitengasser, had lived there since 1511.[2]

In Feuchtwangen, fourteen miles to the east, a monk from the Mönchroth monastery by the name of Johann von Wald, was invited by the local bailiff, mayor, and parish to hold forth on the evangelism in 1523.[3]

The preachers who appeared in the southern part of Hohenlohe near Schwäbisch-Hall were either graduates of Heidelberg or friends of the foundation preacher in Hall, Martin Brenz, or both. By 1525, Brenz had begun to recommend friends as ministers for the surrounding communities, thereby widening the sphere of influence beyond that of Schwäbisch-Hall's sovereignty.[4] Thus, Brenz recommended one of his colleagues, Mathäus Chyträus to Ingelfingen, following the request of its magistracy. Upon completing his education in Tübingen,[5] Chyträus spent some time with Brenz in Hall.[6] He left Ingelfingen to become preacher in Menzingen (Kraichgau) to escape the restraints of the Hohenlohe authorities.

In Reinsberg, the minister Johann Herolt also depended on Schwäbisch-Hall for inspiration and moral support. Being a resident preacher, his case has already been discussed. In Rieden, peasants asked the Hall city government to provide them with a parson.[7] The village already had a chaplain who had been installed and paid by the city. When mass was discontinued in Hall's St. Michael in 1525, the same was ordered for Rieden. From then on, Nicholas Trabant from Hall volunteered to read the sermon in Rieden every second

---

[1] Sigel, Das evangelische Württemberg, Vol. VI, No. 1, p. 205.

[2] Ibid., Vol. VI, No. 2, pp. 755-66; and Beschreibung des Oberamts Crailsheim (Stuttgart: Kohlhammer, 1884), p. 419.

[3] Anton Steichele, Das Bistum Augsburg (Augsburg, 1864), III, 566 and 581.

[4] Friedrich Kantzenbach, "Theologie und Gemeinde bei Johannes Brenz," Blätter für Württembergische Kirchengeschichte, LXV (1965), 34.

[5] Mathäus Chyträus is not mentioned in the Tübingen Matrikeln. According to Hartmann, in Johann Brenz: Leben und ausgewählte Schriften, I, 188, and Sigel, Das evangelische Württemberg, X, 188, he was a Tübingen student, however.

[6] Hartmann and Jäger, Johannes Brenz, I, 188.

[7] Gmelin, Hällische Geschichte, pp. 742-43.

Sunday.[1]

Main Basin

Several towns and villages along the Main river hosted out-of-town preachers. Miltenberg was one of the few communities in the prince-bishopric Mainz that had a Lutheran minister as early as 1522.[2] Dr. Johannes Draconites, also called Trach or Carlstadt, had studied under Erasmus in the Netherlands, then under Luther in Erfurt and Wittenberg.[3] His relative, the bailiff Friedrich Weygand of the Mainz Electorate obtained the office in Wertheim for him.[4] Trach led the Miltenbergers to openly rebel against their territorial ruler, but was forced to leave the town in 1523. He sought asylum in Wertheim to help Franz Kolb in his work. During the same year he returned to Wittenberg, however, to finish his doctorate in theology.[5]

Wertheim, the capital of the county with the same name, was governed by Count Georg II. He had been a student of Luther's in Wittenberg in 1520, and occasionally exchanged letters with him. When the count asked Luther for a preacher two years later, the latter sent Jakob Strauss, an ebullient sermonizer. However, Count Georg soon dismissed him because of his radical approach.[6] At the request of the secular ruler, Franz Kolb took over Strauss' position. Kolb, once a humanistically inclined monk (Carthusian), now read the mass in German. Kolb was born outside of Basle and resided mostly in Freiburg and Berne, Switzerland. With the outbreak of the religious dispute, he joined the Lutheran, and later the Zwinglian movement. This was to cost him his preaching office in Freiburg. He thence went to Nuremberg and met the

---

[1] Sigel, Das evangelische Württemberg, Vol. VI, No. 2, p. 670. Trabant formerly was preacher in Flehingen. Brenz invited him to Schwäbisch-Hall.

[2] Albrecht Otto, Die evangelische Gemeinde Miltenberg und ihr erster Prediger, Schriften für das deutsche Volk (Halle, 1891), p. 8.

[3] Neu, Pfarrerbuch der evangelischen Kirche, II, 120.

[4] G. Kawerau, "Johannes Draconites aus Carlstadt," Beiträge zur Bayerischen Kirchengeschichte, III (1897), 250.

[5] Ibid., p. 258.

[6] Wilhelm Störmer, "Obrigkeit und evangelische Bewegung: Ein Kapitel fränkischer Landes- und Kirchengeschichte," Würzburger Diözesan-Geschichtsblätter, XXXIV (1972), 115.

ruler of Wertheim during the Imperial Diet in 1522.[1] If it is true that Luther personally suggested Kolb to Georg II, it seems unlikely that he would have sided with Zwingli and his followers. But as early as 1524, the former Carthusian criticized Luther for his conservative stand and drew up a Zwinglian church order.[2] He left Nuremberg at the end of 1524, possibly on order from Georg II, since Kolb's chaplain was also dismissed some years later for similar reasons.[3] Both Strauss and Kolb lived in Wertheim at the expense of Count Georg.[4]

The earliest registered date for Eberlin's appearance in Wertheim is early July, 1526. He was the Prädikant, or preacher rather than the parson there.[5] In importance he excelled the latter instead of being subordinate to him. Eberlin came from Ansbach and introduced a church order which Luther personally had drawn up for him. The exact year of the inauguration is not known, although it must have occurred between 1526 and 1530.[6] However, the people in and around Wertheim, like the rural population in the hinterland of Lindau, opposed secession from the Roman Catholic faith.[7]

In Bettingen, Dertingen, Nassig, Niklashausen, and Wenkheim, the princes of Wertheim, Löwenstein-Wertheim-Freudenberg, and Rosenberg, jointly held patronage over the churches. By 1530, all the vicinity of Wertheim had

---

[1] Heinrich Neu, Geschichte der evangelischen Kirche in der Grafschaft Wertheim (Heidelberg: Winters, 1903), p. 9.

[2] See Matthias Simon, "Kirchenordnungen Frankens," in Die evangelische Kirchenordnung des 16. Jahrhunderts, ed. Emil Sehling (Tübingen: Mohr, 1963), pp. 124-57.

[3] Matthias Simon, "Zur Reformationsgeschichte der Grafschaft Wertheim," Zeitschrift für Bayerische Kirchengeschichte, XXIX (1960), 125, footn. 32. In 1525 Kolb went to Nuremberg. He left in 1527 because of his pro-Zwinglian bias. See Köhler, Luther and Zwingli, I, 230-31.

[4] Simon, "Zur Reformationsgeschichte der Grafschaft Wertheim," p. 126.

[5] Vierordt's claim that Georg II appointed Eberlin as parson in 1525 seems hardly correct. See Vierordt, Geschichte der Reformation im Grossherzogtum Baden, I, 122.

[6] Fritz Kobe, Die erste lutherische Kirchenordnung in der Grafschaft Wertheim, Veröffentlichungen des Vereins für Kirchengeschichte in der evangelischen Landeskirche Badens (Lahr: Moritz Schauenburg, 1933), p. 9.

[7] Alexander Kaufmann, "Einige Bemerkungen über die Zustände des Landvolks in der Grafschaft Wertheim während des 16. und 17. Jahrhunderts," Freiburger Diözesan-Archiv, II (1866), 53.

a Protestant church service, in spite of the strong opposition of the prince-bishop of Mainz and Würzburg.[1]

Johann Selzach filled the preachership in nearby Bettingen in 1529. It is possible that Selzach was still under the direction of the Protestant community of Wertheim during the interval of nine months or longer which elapsed between Kolb's departure and Eberlin's arrival. Thus, Eberlin would have been partly responsible for restoring mass, although in a Protestant form. It had been strictly abolished by his predecessor. In Dertingen, Jakob Wendt also preached the Lutheran way beginning in 1528,[2] while Leonhard Knetzer was the Protestant preacher in Niklashausen in the Tauber valley, after 1529.[3] By the year 1530, seventeen preachers were engaged in holding Protestant sermons in twenty-two villages in the Wertheim vicinity.[4] The waves of Eberlin's reform attempts were felt also in the Archbishopric of Mainz to the south. In a church trial that took place in 1526 against the married preacher of Uissigheim, a witness laid the blame on "Wertheim's clergy and other ministers that have married."[5]

The Reformation movement, to the contrary, had but little effect in the vicinity of Würzburg and further upstream. The bishop of the Würzburg diocese successfully prevented the recruitment of Lutheran preachers. Nevertheless, groups were forming that showed hatred and aversion toward the bishop. Still further up the river from the bishop's seat, in Kitzingen, Dettelbach, and Volkach, early signs of reformatory activities began to be felt also. From there they spread to several localities within the Würzburg bishopric.[6] For example, Kitzingen, an important Main-bridgetown on the road between Würzburg and Nuremberg, had young Christoph Hoffmann from Wittenberg take over the preaching foundation in 1522.[7] Hoffmann was encouraged by Schenk von

---

[1] Neu, Pfarrerbuch der evangelischen Kirche Badens, I, 301.

[2] Ibid., I, 303.   [3] Ibid., I, 305.

[4] See Heinrich Neu, Geschichte der evangelischen Kirche in der Grafschaft Wertheim (Heidelberg: Winter's, 1903), p. 18, footnote.

[5] Andreas Veit, Kirche und Kirchenreform in der Erzdiözese Mainz im Mittelalter der Glaubensspaltung und der beginnenden tridentinischen Reformation (1517-1618), Erläuterungen und Ergänzungen zu Janssens Geschichte des deutschen Volkes, No. 10, 3 (Freiburg: Herder, 1920), p. 18.

[6] Lothar Michel, Der Gang der Reformation in Franken: Auf Grund kritischer Übersicht über die bisherige Literatur dargestellt, Erlanger Abhandlungen zur mittleren und neueren Geschichte, No. 4 (Erlangen: Palm & Enke, 1930), p. 7.

[7] Simon, Evangelische Kirchengeschichte Bayerns, p. 169.

Sieman. In spring of 1524, the former Benedictine monk Diepold Peringer also arrived from the Nuremberg area to deliver a sermon on Corpus Christi Day. Peringer was expelled at the order of Margrave Casimir, who suspected him of being an adherent of Carlstadt.[1] Peringer moved to nearby Rothenburg.[2] However, Johann Meglin from Ansbach boosted the reformatory process by officially introducing German church hymns and communion in both forms (<u>sub utraque specie</u>). The migrating preachers Wolfgang Vogel and Diepold Peringer mentioned above, took charge of the three villages Eltersdorf, Thon, and Wöhrd, all north of Nuremberg. They proclaimed the Word there in 1524, after being evicted from Bopfingen and Ulm. In 1527-28, Vogel was replaced by Andreas Althamer, who had been denied his preaching office in Gmünd.

North Baden

Most towns and villages on the upper right side of the Rhine hosted preachers who had come from outside, not unlike the twenty or so small knightly localities in the Kraichgau. Whereas a large percentage of the preachers between Miltenberg, Nuremberg, and Nördlingen had come directly from Wittenberg, the preachers on the upper Rhine were from either Heidelberg, Strasbourg, or Freiburg. The different religious orientation these places had, the former Lutheran, the latter Zwinglian-Bucerian, resulted in the westward expansion within Württemberg of these two evangelical faiths. No doubt, the Lutherans were at a disadvantage on this open flank between Basle and Frankfurt, since they lacked a strong common diffusion base from which they could spread their ideology.

The spatial extension of the reformed church north of Frankfurt and Hesse, as well as into the Palatinate, is due partly to the naturally favorable location of Strasbourg in the Rhine basin as well as the practical orientation of its reformers.

In the Palatinate town of Bretten and its environs, the Reformation did not break ground until shortly prior to 1534.[3] The first Protestant preacher, Johann Eisenmayer (or Isenmann, <u>Siderocrates</u> in Latin), came from neighboring

---

[1] In numerous vigorous and popular sermons, Peringer advocated the Lutheran theses, but rejected the saint's images as "painted idols." Paul Kalkoff, p. 129.

[2] Ernst Enders, <u>Dr. Martin Luther's Briefwechsel</u> (Calw: Ev. Verein, 1893), V, 154.

[3] Neu, <u>Pfarrerbuch der evangelischen Kirche Badens</u>, I, 32.

Kürnbach. He had become familiar with Luther's theology in Weinsberg, where he was associated with the local preachership held by Erhard Schnepf. Both were forced to leave Weinsberg in 1522, with Schnepf going to Guttenberg, and later to Wimpfen.

In Jöhlingen, located between Bruchsal and Durlach, a Lutheran preacher began sermonizing in 1527.[1] The villagers were probably already familiar with Luther's teachings through Anton Engelbrecht in Bruchsal. The name of the parson is not known.

By 1520 the Reformation in Ettlingen had already gained a considerable number of adherents. In the spring of the preceding year, a priest of Ettlingen had written to Luther, zealously endorsing his view on the indulgence issue. The sender of the letter might have been Irenicus, who was born and raised there. Irenicus had enrolled in Tübingen in 1516, but transferred to the university in Heidelberg the following year. There he became a member of the academic senate. His participation at the disputation in April 1518, as well as his affiliation with men like Bucer, Brenz, Billican, Melanchthon, and other important figures of the Reformation made him a determined follower of Luther's creed.[2] There is no proof that Irenicus returned to Ettlingen in 1524 or shortly before.[3] All available evidence suggests that he became foundation preacher at the court of Margrave Philipp I at Baden-Baden. There he probably got married before accompanying the Margrave to Esslingen.[4]

In Irenicus' place, Leonhard Weller became Ettlingen's preacher in October, 1522.[5] Weller entered the University of Heidelberg in July, 1515, received a B.A. in 1516,[6] and an M.A. in 1519.[7] Whether he participated at Luther's debate is doubtful. Nevertheless, the pro-Lutheran attitude of his followers at the university must have left an imprint on Weller. It would not be sur-

---

[1] Ibid., p. 46.

[2] See Horawitz, "Franz Irenicus," Allgemeine Deutsche Biographie, Vol. 18, p. 583.

[3] Vierordt asserts that Irenicus began to preach in Ettlingen in 1524 at the earliest.

[4] See p. 86.

[5] Kattermann, Die Kirchenpolitik Markgraf Philipp I, p. 20, footn. 66.

[6] Gustav Töpke, Die Matrikel der Universität Heidelberg (Heidelberg: Winter's, 1903), I, 503.

[7] Ibid., p. 439.

prising if Irenicus recommended this capable student for the pastorate in Ettlingen. Weller occupied it for almost ten years, during which he married a local burgher's daughter. When Margrave Philipp, under the influence of King Ferdinand, reversed his earlier decision and announced his opposition to the newly emerging Lutheran church, Weller left the area to become preacher in Gemmingen. Ambrosius Blarer tried to win Weller for Esslingen in 1532; however, it seems that he was unwilling to shift allegiance to the reformed body of the new faith.[1]

In Grötzingen, Christoph Sigel, highly esteemed because of his intelligence and gentle character, preached several years before he left the town in 1532. His reformatory-minded sympathizers in Strasbourg found him a new position in Zweibrücken. In spring, 1533, he came to Ulm and asked to be transferred to Überkingen, a neighboring spa.[2]

Strasbourg Area

Numerous localities along the right side of the Rhine opposite Strasbourg, as well as the cities in the Kinzig Valley, lay in the sphere of influence of this large Alsatian metropolis. It could boast of renowned church reformers and organizers such as Martin Bucer, Caspar Hedio, Jakob Sturm, and Wolfgang Capito. Most of the Ortenau nobility, including its abbots, were citizens of Strasbourg, as was noted earlier in reference to several Ortenau knights and Count Wilhelm of Fürstenberg. Inhabitants of this economic, cultural, political-administrative, and ecclesiastical central place were always informed on reformatory changes. It is, therefore, not surprising that tributary towns and villages addressed themselves to the reformers of this focal center whenever their citizens demanded a preacher of the new variety.

On the request of the burghers of Kehl and its baron Ludwig Böcklin von Böcklinsau (a naturalized citizen of Strasbourg), margrave Philipp I and the magistracy of Strasbourg appointed Leonhard Volk as its parson. Born in Augsburg, Volk was deacon at Alt St. Peter in Strasbourg until he was transferred across the Rhine to Kehl.[3]

---

[1] According to a letter that Ambrosius Blarer sent Martin Bucer on February 29, 1532. See Traugott Schiess, ed., Briefwechsel der Brüder Ambrosius und Thomas Blarer 1509-1548 (Freiburg: Fehsenfeld, 1908), I, 327-28.

[2] Gustav Bossert, "Badisch-Pfälzische Reformationsgeschichte," Zeitschrift für die Geschichte des Oberrheins, XIX (1904), 46. In fall of 1534 Sigel accepted an invitation from Esslingen, where he stayed until his death in 1542.

[3] Neu, Pfarrerbuch der evangelischen Kirche Badens, II, 631.

Six miles north of Kehl, the community of Hanau hired a Protestant preacher probably as early as 1525, and, in a letter to the Strasbourg magistracy aired their "imploring request . . . not to forsake them, but to remember what an offense it would be, to be forced to listen once more to a priest who condemns the Holy Gospel." Nevertheless, in 1527, the Strasbourg bishop forced Georg Wickenhauer, the evangelical minister to leave the village.[1] In neighboring Freistett, an old fishing village, reformatory efforts starting as early as 1523 can also be traced back to Strasbourg.[2] However, whether Freistett had an official preacher prior to the conversion of the count of Hanau is unknown. Lichtenau maintained an evangelical preacher in late January or early February of 1525. He was Martin Enderlin, the chaplain of Rudolf von Baden in Strasbourg. Enderlin acquired Strasbourg citizenship and severed all ties to the Roman-Catholic Church by marrying a Strasbourg girl in October 1523. When Lichtenau asked Strasbourg for a Protestant preacher, the magistracy dispatched Enderlin to that community. The bailiffs of Hanau and Bitsch arrested him for his religious viewpoint, although Strasbourg later obtained his release.[3] The dominion of the Catholic Church had ended with the Peasants' War, and after April 1525, Lichtenau had nothing but Protestant preachers.[4]

Communities in the Kinzig Valley such as Gengenbach, Offenburg, Biberach, and a number of mining centers to the north, also were influenced by the example of Strasbourg in their decision to embrace the Reformation. Count Wilhelm of Fürstenberg asserted in 1542 that he took no part in introducing the new religion in his territory. Instead, he claimed, it evolved naturally and independently of his efforts.[5] Count Wilhelm was born in 1492 in Haslach and educated at the universities in Freiburg and Strasbourg. His restless and fiery character seemed to have been molded particularly by Franz of Sickingen and the close relationship he established with Jakob Sturm and Caspar Hedio during his frequent stays in Strasbourg. He openly sided with the new faith in 1529, when he accompanied the Upper German reformers Zwingli, Oecolompadius, Bucer,

---

[1] Vierordt, Geschichte der Reformation im Grossherzogtum Baden, p. 251.

[2] Erich Keyser, ed., Badisches Städtebuch (Stuttgart: Kohlhammer, 1959), p. 235.

[3] Ludwig Lauppe, "Die Reformation im klösterlich-schwarzwäldischen Kirchspiel Scherzheim-Lichtenau," Die Ortenau, 1952, pp. 74-75.

[4] Keyser, Badisches Städtebuch, p. 295.

[5] Gothein, Wirtschaftsgeschichte des Schwarzwaldes, I, 273-74.

and Hedio on their homeward journey from Marburg to Strasbourg.[1] An enemy of monastic orders and the high clergy, he showed great interest in the ecclesiastical changes initiated in Strasbourg, although he never formally changed his religious confession.[2]

A large proportion of the citizenry of the imperial town of Gengenbach were small craftsmen and wine growers who had good connections with the adjacent metropolis.[3] Since the local monks had relinquished their right of appointment in 1437, the town created its own benefice, which was connected with the lay patronage. As counterpart to the local monastery, this benefice was especially susceptible to public pressure in favor of reformatory changes beginning in 1523.[4] It seems likely that Martin Bucer preached in Gengenbach that year, since he had no parish of his own before spring 1524. Konrad Knecht (Servitor), the first regular Protestant preacher, began holding sermons toward the end of 1524.[5] Unfortunately, very little is known about his background or activities; his hometown probably was in Vorarlberg or Upper Swabia.

Lucius Kyber, one of three Gengenbach preachers and draftee of the local church order in the year 1538, must have been instrumental in furthering the Reformation movement as early as 1525.[6] He came from the Austrian town of Bludenz in Vorarlberg, where he served as chaplain at the hospital. Kyber and Thomas Gassner, "Lindau's evangelists" were probably friends from the Latin school in Feldkirch[7] and had become early champions of the Reformation in

---

[1] Ibid., p. 309.

[2] Ernst Münch, Geschichte des Hauses und Landes Fürstenberg (Aachen: Meyer, 1829), p. 20.

[3] Willy Andreas, "600 Jahre Reichsstadt Gengenbach," Zeitschrift für die Geschichte des Oberrheins, CVIII (1960), 300.

[4] Krebs, "Politische und kirchliche Geschichte der Ortenau," p. 123.

[5] Karl Bender, "Die Reformation in Gengenbach," in Beiträge zur Badischen Kirchengeschichte, Veröffentlichungen für Kirchengeschichte in der evangelischen Landeskirche in Baden, XXII (1962), 11-12.

[6] Since the reformer served only until 1523 in Bludenz, and was superseded by Gassner in 1524, it might well be that Kyber arrived in Gengenbach prior to the latter date. Also, one of Kyber's sons was born in Gengenbach in 1525.

[7] Ernst-Wilhelm Kohls, Evangelische Bewegung und Kirchenordnung: Studien und Quellen zur Reformationsgeschichte der Reichsstadt Gengenbach, Veröffentlichungen des Vereins für Kirchengeschichte in der evangelischen Landeskirche Baden, No. 25 (1966), p. 19. Ernst Staehelin states that Kyber was active in Gengenbach "in 1525 or earlier." See "Oecolompadiana," Basler Zeitschrift für Geschichte und Altertumskunde, LXV (1965), 176.

Bludenz.[1]

The evangelical lay priest (Laienpriester) who was appointed in 1526 faced firm opposition from the local convent. But the delegates of Gengenbach valiantly defended their clergymen before representatives of the Swabian District (Schwäbischer Kreis) that met in Esslingen.[2] The complaints lodged against Kyber included the following: (1) he rejected the Roman-Catholic principle of transsubstantiation in communion;[3] (2) he and his aides abolished old customs during baptism, which they began to perform in German; (3) he adopted the evangelical hymns that were sung in Strasbourg; and (4) he abandoned High Mass and asked his parishioners to help him abolish the "impious" Low Mass as well.[4]

At the demand of Bishop Wilhelm of Strasbourg in early 1527, that the lay priest cease his evangelic activity by abandoning his position, the town council stood firmly behind its parson. The citizens appealed to the imperial mandate established during the Speyer Diet in 1526, which proclaimed the virtue of "preaching the gospel and pure and uncorrupted Word of God everywhere."[5] Gengenbach represented this position even after the Augsburg Diet in 1530.[6] In 1532, Mathäus Erb arrived in Strasbourg from Switzerland to become schoolmaster and reformer of Gengenbach.[7]

Meanwhile, the Reformation spread further up the Kinzig Valley. As early as 1527, the inhabitants of Biberach hired an evangelical preacher. Although they did this in accordance with Zell-Harmersbach, an imperial town, they had not asked the bishop for permission.[8]

---

[1] Burmeister, Thomas Gassner: Ein Beitrag zur Geschichte der Reformation und des Humanismus in Lindau, p. 11.

[2] Ernst Batzer, "Neues über die Reformation in der Landvogtei Ortenau sowie in den Städten Gengenbach und Offenburg," Zeitschrift für die Geschichte des Oberrheins, N. F., XXXIX (1926), 65.

[3] Kyber likewise rejected the Lutheran idea that the body of Christ is in the bread; he believed that the bread was only a token of Christ's body. Ibid., p. 66.

[4] Ibid.   [5] Ibid.

[6] Krebs, "Politische und kirchliche Geschichte der Ortenau," p. 137.

[7] Ibid.

[8] Gothein, Wirtschaftsgeschichte des Schwarzwaldes, I, 270.

The Breisgau

In this Austrian principality in southern Baden a number of towns fell under the influence of Protestant preachers. They were part of the evangelical movement in Freiburg, until government action forced them to take refuge in Strasbourg.

Jakob Otter of Alsace preached in two parishes, Wolfenweiler and Kenzingen simultaneously. Between 1510 and 1520, Otter became a student of theology and a teacher of philosophy in Freiburg. While earning various theological degrees,[1] he became a close associate of Ulrich Zasius, the famous professor of jurisprudence. He immediately supported Luther's teachings as soon as they circulated among Freiburg's humanists in 1518, and by 1520 had earned himself a reputation for being a "die-hard Lutheran" (Erasmi Lutherique adiuratus cliens).[2]

In the village of Wolfenweiler southwest of Freiburg, Otter's predecessor Johannes Kess was accused of pro-Lutheran sentiments in 1519.[3] In spite of his strong commitment to Lutheranism, Otter preached to the Wolfenweilers in a moderate way. Early in 1522, the magistracy in Kenzingen appointed him as the assistant of the Stadtpfarrrektor (rector of the parish) in Kenzingen. Located as they were along the heavily-travelled road between Frankfurt and Freiburg, the population of Kenzingen had already become familiar with Luther's teachings. Otter merely added fuel to a passionate movement that eventually sparked opposition from the ecclesiastical and secular authorities, the Bishop of Strasbourg, and the Austrian government of Breisgau.

In all this, Otter was supported by the magistracy as well as the citizens of Kenzingen. But outside pressure from the Austrian government at Ensisheim finally caused him to depart for Strasbourg. A substantial sector of the town's population were willing to leave and follow him. Only after lengthy negotiations between the Protestant Strasbourg city government and the margrave on the one side, and the Austrian government on the other, was the Protestant group allowed to return. Condemned as an "arch-heretic" for trying to introduce communion under both forms, the town clerk was decapitated and burned, while seven burghers were jailed in Ensisheim.[4] Since any reformatory at-

---

[1] According to the Protocollum proclamacionum of Freiburg's archbishop on February 19, 1519, Otter had earned a licentiate in the Holy Scripture. See Peter Albert, "Die reformatorische Bewegung zu Freiburg bis zum Jahre 1525," p. 15, footnote 1.

[2] Ibid., p. 12.   [3] Ibid., pp. 13-14.

[4] Hans W. Rohde, "Evangelische Bewegung und Katholische Restauration

tempts were almost extinct in Freiburg itself, the guidance for Kenzingen's Protestants must have come from Strasbourg.[1]

In Ehrenstetten, which belonged to the Teutonic order, Ulrich Weber, in an oath of truce admitted that he had invited a Lutheran preacher who "rejected the Mass, the saints, the prohibition of meat during fasting period," as well as other statutes of the church.[2] We do not know who persuaded Weber, nor do we know the preacher who spread this message. The former might have had contact with Peter Spengler, then parson in Schlatt and dean of the Breisach rural chapter.[3]

A former Carthusian monk, Otto Brunfels became Protestant preacher in Neuenburg/Rhine after having stopped there on his way to Switzerland in 1522. Brunfels held an advanced degree from Mainz University and was a friend of Nikolaus Gerbel, Ulrich of Hutten, and Franz of Sickingen. He left Neuenburg before the Austrian government initiated measures against him and his supporters.[4] His successor, Alexander Ryschacher catered to this same group before he was forced to leave Breisach in 1535. Known as the former plebanus of Niederrotweil, Ryschacher was blamed for stirring up reformatory issues.[5] He was given the Neuenburg pastorate only on the condition that he completely disavow "Lutheran, Zwinglian, and other damnable teachings."[6]

Summary

The following conclusions can be drawn from the preceding discussion:

(1) Immigrant preachers appeared mostly in smaller communities and vil-

---

im österreichischen Breisgau" (unpublished Ph.D. dissertation, University of Freiburg/Br., 1957), p. 37.

[1] The peasant unrest in the Kenzingen vicinity in 1513 and later was also caused by Alsatia. See Hans Virck, Politische Korrespondenz der Stadt Strassburg im Zeitalter der Reformation, I (1517-1530) (Strassburg: Trübner, 1882), 104.

[2] Albert, "Die reformatorische Bewegung zu Freiburg bis zum Jahre 1525," p. 74.

[3] Vierordt, Geschichte der Reformation im Grossherzogtum Baden, I, 279-80. Spengler was executed for his attempts to convert the clergy of his chapter.

[4] Brunfels returned to Strasbourg and then went to Berne, where he died as city physician in 1534. See F. W. E. Roth, "Otto Brunfels," Zeitschrift für die Geschichte des Oberrheins, IX (1894), 320.

[5] Albert, p. 74.   [6] Ibid.

lages. These relied mainly on larger cities that had a reservoir of qualified personnel, notably highly qualified preachers.

(2) Where political authorities forced preachers to leave their communities, the clergy headed for the next large city, from which they again re-distributed to smaller localities that demanded their service. Strasbourg, Ulm, Zurich, and Schwäbisch-Hall are the better known re-distribution centers.

(3) From 1525 onwards, outstanding preachers and theologians such as Ambrosius Blarer, Oecolompadius, Bucer, and Brenz visited smaller cities and towns to solidify the Protestant position there.

(4) Cities often hosted an indigenous preacher first who was later replaced by an immigrant preacher. This was the case in Esslingen, Memmingen, Nördlingen, and Riedlingen. Only rarely did a city exclusively rely on immigrant preachers, an exception being Gmünd.

(5) Cities and villages located along important commercial routes had more contact with travelling salesmen who frequently talked about religious subject matters. They were therefore more inclined to adopt the new faith than more isolated communities.

(6) Where an immigrant preacher with a set mind had arrived, he often proved to be an attraction to surrounding villages that were too distant from a primary or secondary center.

(7) Only two preachers were natives of the town in which they preached after their arrival. They were Caspar Kantz of Nördlingen, and Adam Weiss of Crailsheim.

CHAPTER VII

PREACHING FOUNDATIONS (PRÄDIKATUREN) AND
THE DIFFUSION OF THE REFORMATION

The foundation preachers (Prädikanten)[1] belonged in the class of secular clergy (clerici saeculares). In accord with the demands set forth in its bequest, the foundation or preachership was reserved for theologians endowed with a special and superior education at one of the universities. Owing to this and their extensive knowledge of the Holy Scripture, the reputation of a foundation preacher far exceeded that of the regular parson with whom he served in a particular community. This proved to be especially the case in areas and localities where the regular clergy remained indifferent to the religious needs of its congregation. Moreover, one can distinguish three types of preaching foundations: (1) regular, (2) court, and (3) cathedral preaching foundation. They originated in Strasbourg, spread down the Rhine and up the Neckar valleys. Most were found concentrated along the Neckar and Danube rivers (Fig. 16).

Luther's special emphasis on the sermon over the other elements of religious care served to further strengthen the position of the preacher over the parson. The regular parson was responsible for the cura animarum or care of souls, which entailed administering the sacraments and praying for the dead as well as the founder from whom he earned his living. Since sermons did not reap any profit, the creation of preaching foundations meant no financial loss to him. Besides, these foundations often were the main avenue through which reform ideas reached the broad public, since the task of their preachers was to satisfy the popular desire for the spreading and interpretation of the Divine Word (officium praedicationis et lectionis).[2] The transitory nature of sermons given by Protestant migrant and clandestine preachers lacked the enduring quality and effect of those given by the foundation preacher. The layman's modus operandi

---

[1] Various titles were applied to the occupant of a preaching foundation: praedicator, consionator or ecclesiastes, semini verbius (word sower), or simply sator (sower).

[2] Theodor Kolde, D. Johann Teuschlein und der erste Reformationsversuch in Rothenburg o. d. T. (Erlangen: Böhme, 1901), p. 4.

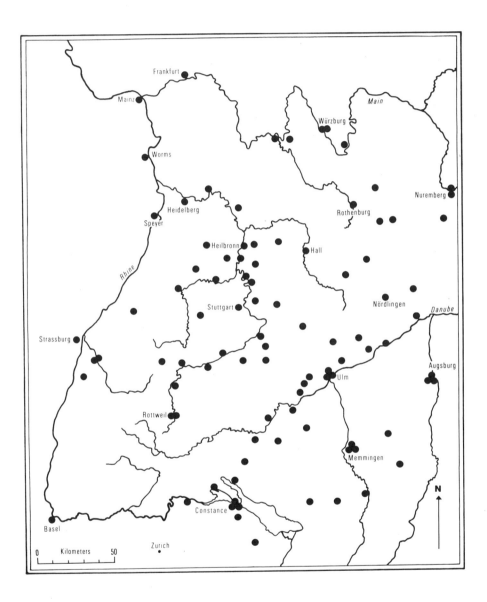

Fig. 16.--Distribution of Preaching Foundations at about 1520

of word-by-mouth was liable to adulterate Luther's ideas with socio-economic and political demands, if not Schwärmer sentiments.[1] Hence it was the continued prudent proclamation to the contrary by foundation preachers that prepared parishes and cities for eventual changes in church service and ecclesiastical organization. It would be very interesting to analyze the sermon texts existing from this period to see in what manner they disseminated the ideas of Luther, Zwingli, and other reformers among the people. No doubt, every city and town had its indigenous religious ways; and it should be possible to draw regional generalizations on the basis of existing similarities and differences. However, this would require a separate line of research that would exceed the already broad limits of the present study.

The existence of a preaching foundation did not automatically guarantee the acceptance of Luther's ideas. To a large degree this depended on the institution or person that filled the position, and whether the founder belonged to the ecclesiastical or secular category. The ecclesiastical corporations, which in the majority of cases had the right to nominate the preacher, were not inclined to select Lutherans; wherever possible, they blocked the effort to spread the new faith. The majority of monks, nuns, prebendaries and provosts who did the conferring, were holders and beneficiaries of the traditional ecclesiastical system. In contrast, the preacherships that were established by city or town were particularly susceptible to the new faith, as is illustrated by the examples of Schwäbisch-Hall, Reutlingen, Giengen, Rothenburg, Kitzingen, Nuremberg, Miltenberg, Kempten, Heilbronn, Dinkelsbühl, Offenburg, and Nördlingen. The importance of the fact that such cities assumed responsibility for hiring and reimbursing the foundation preachers cannot be overlooked. The rights of the city and the magistracy exceeded that of the patron, as the city government had both the right to nominate and present a preacher. No ecclesiastical authority was involved in this process. The magistracy also supervised the appointed preacher in order that he might fulfill his duties as prescribed in the original contract.[2]

Often an outlying monastery was in charge of caring for the souls of urban residents. Since this task was often taken very lightly, the cities were forced

---

[1] Schwärmer or Schwarmgeister ("emotional enthusiasts") was used by Luther to refer to the Anabaptists as well as the followers of Andräas Carlstadt and Thomas Münzer.

[2] Hermann Kienzle, "Rechtliche Grundlagen und Vorrausetzungen der Reformation in Heilbronn" (unpublished Ph.D. dissertation, University of Tübingen, 1921), p. 43.

to seek alternatives. When one considers the large amounts of money and values in kind flowing from the city to a small number of highly privileged clergy, and the often weak yield in spiritual edification offered in return, a negative assessment of costs and benefits must have prevailed.

Why were the preaching positions so open to new ideas? Unlike the benefices and the administration of the sacraments, the sermon and preaching office were not a rigid component of the traditional Roman-Catholic ecclesiastical system. The efforts on the part of the magistracies to gain control over the staffing of these preacherships existed well before the Reformation movement had begun.[1] As it was, the relationship between foundation preachers and the established church was anything but harmonious. Many thought it necessary to attack the clergy and even the bishop from the pulpit, either to upgrade themselves or to win the approval of their parish members. Not a few clergymen seem to have been in constant opposition to the official Roman Catholic Church; and in fact, the Lutheran preachers were the direct outgrowth of this dissident movement.[2]

The importance that was now being attached to sermons in Protestant church service, together with the social prestige of the persons who occupied the preaching offices, made these men extremely influential. Their authority and power went far beyond the rank assigned to them by canon law. According to its statute, the foundation preacher was a chaplain (vicar cooperator) or "fellow" of the parson, who could act only on instruction from the latter. The Reformation movement altogether reversed this relationship. From now on the foundation preacher assumed the role of an organizer and adviser to the magistracy on religious policy in addition to being the obvious opinion leader of the masses. The course of the Reformation, and no less the Peasants' Rebellion, often depended on the attitude of the preacher, and the parson now definitely was thrust into the background.[3]

In order to fulfill his role adequately, it was necessary that the foundation preacher had no duties other than preaching in the vernacular, so that he could

---

[1] Rudolf Herrmann, "Die Prediger im ausgehenden Mittelalter und ihre Bedeutung für die Einführung der Reformation im Ernestinischen Thüringen," Beiträge zur Thüringischen Kirchengeschichte, I (1929-1931), 64.

[2] Albert Braun, Der Klerus des Bistums Konstanz im Ausgang des Mittelalters, Vorreformationsgeschichtliche Forschungen, No. 14 (Münster: Aschendorffschen, 1938), p. 138.

[3] Hermann, "Die Prediger im ausgehenden Mittelalter," p. 64.

"administer the pulpit with diligence."[1] Although it seems that holding sermons was the task of the parson, this duty was frequently neglected due to his inadequate educational background, resulting in a corresponding inability of the majority of parsons to preach.[2] Unlike that of most preachers, the standard education of the average parson allowed only the reading of mass and the chanting of the vigil.[3]

Most foundation preachers were assigned to their position for only one year. In case of a vacancy local clergymen informed their friends in other cities and towns about it.[4] The applicants were screened by the patron of the foundation. Sometimes the professors at one or several universities were informed regarding a vacancy in order to let them suggest a qualified person. The role of nominating a candidate usually fell to the university which was closest and therefore attended by the majority of the town's students. The examples of Schwäbisch-Hall, Weinsberg, and Heilbronn applying to the university in Heidelberg illustrates this trend. Prior to the selection[5] of a candidate by the donor, the applicant was asked to give one, and sometimes several trial sermons before the local congregation.[6]

Generally, foundation preachers delivered their sermons from a pulpit or "preaching chair"; as a rule, the gospel or epistle of the day served as the topic. Delivered in a simple manner, it aimed at improving the morality of the congregation.[7]

---

[1] Otto Kähni, "Reformation und Gegenreformation in der Reichsstadt Offenburg," Die Ortenau, XXX (1950), 25.

[2] Julius Rauscher, "Die Prädikaturen in Württemberg vor der Reformation: Ein Beitrag zur Predigt- und Pfründengeschichte am Ausgang des Mittelalters," Württembergisches Jahrbuch, No. 2 (1908), p. 153.

[3] Ibid.

[4] In Heilbronn, in November of 1520, a number of persons applied mostly through the syndicus Johann Grienbach, or the magistrate Dr. Johann Linck. The latter suggested the following persons: Dr. Schappeler of Memmingen, Martin Cless of Göppingen, and Leonhard Erbach, later rector of Heidelberg university; Moriz von Rauch, ed., Urkundenbuch der Stadt Heilbronn, 1501-1524, Württembergische Geschichtsquellen, No. 19 (Stuttgart: Kohlhammer, 1916), pp. 567-70.

[5] Rauscher, "Die Prädikaturen in Württemberg," p. 166.

[6] Where the magistracy was the donor, the foundation preacher was elected.

[7] Rauscher, "Die Prädikaturen in Württemberg," p. 170.

The endowment of the benefice consisted mainly of a sum of money that was sometimes supplemented with agricultural produce. Only in a few cases do we know the capital sum of the foundation as well as the annual amount of interest (generally 5 percent) and the annual salary of the preacher.[1] However, it is known that Dornstetten in 1493 had an endowment of only 200 Rhenish guilders, and hence a meager annual salary of 10 guilders; on the other hand, the remuneration at Ellwangen, Brackenheim, Augsburg, Göppingen, Isny, Memmingen, Reutlingen, Rothenburg, Schwabach, Rottweil, Ulm, and Weil der Stadt amounted to 100 guilders.[2] The salaries of the foundation at Dinkelsbühl and Strasbourg were even higher, with no less than 150 and 1,200 guilders, respectively.

South-Franconian Cities

Although this area lacked a metropolis, it did possess three medium-sized cities, namely Schwäbisch-Hall, Heilbronn, and Nördlingen. This region furthermore lay along the east-west axis formed by the roads from the Rhine-Neckar cities of Speyer and Heidelberg with Nuremberg and Donauwörth. Consequently, all sorts of goods, particularly wine, were regularly exported over this network to Nuremberg and other Bavarian cities, whose harsh climate prohibited the successful pursuit of viticulture.

In spite of the political and economic influence emanating from the Duchy of Württemberg to the south, the larger towns were oriented toward the already mentioned communication routes. In this context, inhabitants of Wimpfen and Heilbronn preferred to be identified as Kraichgauers instead of Swabians;[3] peasants in the Kraichgau villages of Flehingen wanted to belong to the Palatinate instead of to Württemberg. A possible explanation of these preferences was the belief that Franconians and Swabians had different ethnic characteristics; the former being regarded as a more relaxed type, wittier and more prone to reli-

---

[1] Ibid., p. 174.

[2] The preaching foundation at St. Leonhard in Stuttgart did not allow the salary to exceed 100 guilders; the excess amount was automatically transferred to the Salve-Regina fellowship in that city.

[3] This sentiment was expressed by Ladislas Suntheim of Ravensburg (Kraichgau) in the following words: "Die von Haylprun und Wympffen wellen nit Swaben sein, aber Krächkeyer." See Eduard Lehmann, "Miszellen zur Geschichte des Kraichgaus," Württembergische Vierteljahreshefte für Landesgeschichte, VII (1884), 127.

gious change. The Alemannic temperament, on the other hand, was thought to be religiously more ardent and enduring, and hence more inclined to cling tenaciously to the old traditional line of thought.

The overwhelming role the university in Heidelberg played in the regional diffusion of the new faith cannot be ignored. Its academic brilliance was felt particularly within a 35-mile radius. Whereas Heilbronn and Weinsberg sent five times as many students to nearby Heidelberg rather than to Tübingen, Dinkelsbühl responded with an equal number to each school (Table 3). These proportions illustrate not only Heidelberg's wide intellectual appeal, but also mirror its commercial status within southern Franconia. We observe here a decrease in communication because of increasing distance.

But the pattern just described fits only the larger towns of Heilbronn, Schwäbisch-Hall, Pforzheim, and Nördlingen. It is a curious fact that smaller towns such as Brackenheim, Weil der Stadt, Ellwangen, and Dinkelsbühl sent the majority of their students to Tübingen. A possible explanation may be their equally good road connections and institutional ties to other Swabian cities. Since the university in Tübingen played no role in the diffusion of the Reformation, the above facts might seem trivial; but it is these and other related ties that partly help to explain the dual influence of the Lutheran and the Zwinglian schools on the Reformation movement in southern Franconia. A description of the Kraichgau and Brackenheim later in this chapter provides additional support for this view.

Schwäbisch-Hall[1]

At the end of the sixteenth century, this imperial city had 1223 households (Herdstätten) and a population of slightly over 7000.[2] One reason for its strong east-west orientation was its location north of the Swabian Forest, a region of low population density. It separated Hall and the County of Hohenlohe from the more densely settled middle Neckar and Rems areas to the southwest. Hall's influence was therefore stronger to the north, east, and west, and it served as a trading and transportation center for a fairly large hinterland aligned in these directions. Owing to the leadership of Johannes Brenz, it also served as subregional diffusion center for the Reformation in southern Franconia. A third

---

[1] Also referred to as Hall.

[2] Erich W. Keyser, ed., Württembergisches Städtebuch (Stuttgart: Kohlhammer, 1962), p. 206.

advantage was that here the main road from Heidelberg to the east split up into several branches that served Ellwangen, Dinkelsbühl, Crailsheim, and Ansbach. Hall's central location was established in the fourteenth century, when the city conferred its charter on the nearby towns of Ingelfingen, Künzelsau, and Crailsheim (Fig. 14).

The city council of Schwäbisch-Hall, like that of Heilbronn and Weinsberg, asked for counsel from Heidelberg university to fill their vacant preachership with a qualified clergyman. Thus, Johannes Brenz, like his predecessor Doldius, was chosen upon the recommendation of the university.[1] Since the majority of Hall's students attended Heidelberg university, this was a reasonable choice. Heilbronn sent five times as many students to Heidelberg between 1515 and 1524 than to Freiburg, Wittenberg, and Tübingen.

The Protestant sympathizers who were responsible for the appointment of Brenz were probably members of the so-called Neupatriziat, or new burgher class that replaced the old patriciate in the magistracy.[2] The families Gräter, Isenmann, and Hofmeister, which were later to become important pillars of the Reformation movement, favored the hiring of an outspoken Lutheran preacher. Hanns Wetzel, a member of the magistracy since 1517 seems to have been instrumental in swaying the city government in favor of Brenz.[3]

In September of 1522, Brenz held his trial sermon before the congregation at St. Michaels, the main parish church of Hall. Isenmann accompanied Brenz to the imperial city. The magistracy must have been satisfied with Brenz' sermon, since he was awarded the office with an annual salary of 80 guilders. Isenmann, on the other hand, received the subordinate position of parson. Brenz' affirmation, however, was not a signal that the Reformation movement was underway.[4] The traditional worship of saints, pilgrimage, and solemn processions that prevailed in the city demanded that Brenz adopt a wait-and-see attitude. His sermons from the year 1523 dealt with the importance of a belief and trust in the eternal Word of God. On St. Jakobi Day, he openly

---

[1] Gustav Bossert, "Zur Brenzbiographie," Blätter für Württembergische Kirchengeschichte, X (1906), 105.

[2] Julius Gmelin, "Hall im Reformationsjahrhundert," Württembergisch Franken, N.F., VII (1900), 9, 12.

[3] Friedrich Wilhelm Kantzenbach, "Theologie und Gemeinde by Johannes Brenz, dem Prediger von Hall," Blätter für Württembergische Kirchengeschichte, LXV (1965), 4.

[4] Ibid., p. 8.

attacked the worship of saints as idolatry.[1] This sermon, which was printed in Ulm, made a strong impression on Wolfgang Rychardus, the humanist and doctor in that city.[2]

In September of 1523, Brenz established contact-by-letter with Adam Weiss of neighboring Crailsheim. No mention of a common bond or important events was made, however, until late in 1525, when Brenz sent Isenmann to Weiss with a long explanation of the Syngramma suevicum. He also made inquiries about the church organization set up by Weiss in Crailsheim. The year 1524 was the decisive year for Brenz and the Reformation. With great difficulty was he able to abolish the "blasphemy" of Corpus Christi Day by replacing it with an evangelic sermon.

On Christmas, 1525, Brenz introduced a plain communion under both forms in St. Michaels. It was probably connected to a sermon-church service (Predigtgottesdienst). The church organization, drawn up in 1526, brought with it the re-emergence of the basic theme of the mass. It is possible that Wittenberg had set the example. On a special request of the Margrave Georg of Brandenburg-Ansbach, Brenz participated at the Marburg religious talks between Luther and Zwingli in 1529. During that same year, Brenz advised the ruler concerning the re-organization of the chapters in Ansbach and Feuchtwangen.[3] He and Weiss had made a written assessment about the state of the chapters and priest brotherhoods. The co-operation between the two reformers increased from this time on and their cordial understanding is spelled out in two letters which Brenz addressed to Weiss in 1532. The first gives information about the state of the Reformation in Heilbronn;[4] the second asks Weiss whether he knows anything about followers of Zwingli in his sphere of influence.[5]

---

[1] Ibid., p. 9.

[2] In all probability it was Martin Frecht in Heidelberg who called Rychardus' attention to Brenz. See Gustav Bossert, "Zur Brenzbiographie," Blätter für Württembergische Kirchengeschichte, X (1906), 107, footnote 1.

[3] Hans-Joachim König, "Die Freundschaft zwischen Johannes Brenz und dem Crailsheimer Pfarrer Adam Weiss," Württembergisch Franken, LV (1971), 89.

[4] Theodor Pressel, Anecdota Brentiana (Tübingen: Heckenhauer, 1868), p. 121.

[5] Ibid., pp. 122-23.

Dinkelsbühl

This town had 749 households, or nearly 5000 inhabitants in 1533.[1] It had a preachership long before the Reformation began, as it was established by the citizens as a countermeasure against an indolent clergy. The first Lutheran preacher here was Conrad Abel (Abelius) from the vicinity of Brackenheim, west of Heilbronn. After attending the univerisites in Heidelberg and Tübingen, he entered a Franciscan monastery in Ulm; he left the order in 1522 when he was appointed foundation preacher by the city of Dinkelsbühl.[2] It seems likely that Abel had close contact with Conrad Sam, who occupied a similar position in Brackenheim with intimate ties to Ulm. Possibly they had met each other in Heidelberg; at any rate, it is likely that Abel's appointment was the result of commercial dealings between Dinkelsbühl and the wine-growing area around Heilbronn.

The Peasants' War resulted in a set-back for the reform-minded Protestants that lasted into the 1530's, when Brenz and Weiss recruited another Lutheran preacher for the imperial city of Dinkelsbühl. The initiative for this came from its former burgomaster, Mathias Rösser and Michael Bauer, the former syndicus and now Kirchenpfleger[3] of the city. Through the agency of Weiss,[4] Brenz offered personally to come to Dinkelsbühl to give services, so that a speedy religious transformation could take place. In spite of their full approval, however, Rösser and Bauer had to decline this offer because of opposition by political rulers and certain members of the magistracy.[5] It may be

---

[1] Erich Keyser, ed., Bayerisches Städtebuch (Stuttgart: Kohlhammer, 1967), I, 150.

[2] Mattias Simon, Pfarrerbuch der Reichsstädte Dinkelsbühl, Schweinfurth, Weissenburg in Bayern und Windsheim, Einzelarbeiten aus der Kirchengeschichte Bayerns, No. 33 (Nuremberg: Verein für bayerische Kirchengeschichte, 1962), p. 6.

[3] These superintendents of churches, convents, monasteries, and charitable institutions in the city were members of the magistracy and appointed by the latter to their position.

[4] Weiss was in close contact with Michael Bauer and Hans Harscher, a Dinkelsbühl innkeeper. Harscher, whose proficiency and integrity was held in high esteem by Weiss, had studied Luther's writings, which he regularly received from Frankfurt. In a letter from 1533, Weiss asked Harscher for an extra copy of a newly published work on Luther's theology, as well as news about the Protestant Frankfurt preachers. See Gustav Bossert, "Briefe und Akten zur Geschichte der fränkischen Reformatoren," Theologische Studien aus Württemberg, VII (1886), 4, 7.

[5] König, "Die Freundschaft zwischen Johannes Brenz," p. 91.

that Brenz' appointment, who was already well-known throughout southwestern Germany, might have caused too strong a reaction among the Catholic opponents. Thereupon, Brenz and Weiss proposed Bernhard Wurzelmann, son of the burgomaster of Wimpfen to be preacher in Dinkelsbühl. He was known to Brenz from their student days at Heidelberg and also through his brother, Maternis Wurzelmann, syndicus in Schwäbisch-Hall since 1530.

The Schwaigern preacher was readily accepted by the magistracy of Dinkelsbühl. Michael Bauer first negotiated with Maternis Wurzelmann, then went to Heilbronn to talk directly to Bernhard Wurzelmann. Three weeks after his recommendation, Bauer informed Brenz, Maternis, and then Weiss about the nomination of Wurzelmann to the office of first Lutheran pastor in Dinkelsbühl. Only one month thereafter he introduced communion under both forms and abolished mass.[1]

Nördlingen

The imperial city of Nördlingen, with over 10,000 residents, was situated in the fertile Ries Basin. It was well-known throughout southern Germany for its annual fair, second only to that in Frankfurt/Main. The return of the city's representative from the Diet in Worms in 1521 led to an increased demand for a Protestant preacher. The university, which most of Nördlingen's sons attended, was Heidelberg; Wittenberg and Freiburg were also well attended, but Tübingen less so. Since the city was located immediately south of the Swabian-Franconian ethnic boundary, it was subject to conflicts arising from the Lutheran and Zwinglian doctrines emanating from Heidelberg and Augsburg-Ulm.

On October 31, 1522, the magistracy of Nördlingen hired Theobald Billican from Weil der Stadt. He was furnished with a preachership at St. George's Church, and he signed a ten-year contract with the city.[2] The occupant was expected to be a Doctor of the Holy Scripture, and the salary was to be no less than 100 guilders annually. Continuing in the same vein as he had in Weil, Billi-

---

[1] Ibid., p. 92.

[2] The Cistercians in Heilbronn, who had the ius praesentandi, were unwilling to follow the demands of Nördlingen's citizens and furnish the preaching foundation. No doubt, the monastery foresaw the results of such an action, although Bishop Stadion of Augsburg saw nothing wrong in conferring the position on Billican. See Zoepfl, Das Bistum Augsburg und seine Bischöfe im Reformationsjahrhundert, II, 43; Max Cantz, "Caspar Kantz und die Nördlinger Reformation," 12. Jahrbuch des Historischen Vereins für Nördlingen (1928), p. 166.

can preached the gospel strictly. He rejected the traditional view of the mass as an offering and slandered the mass priests as a "bunch of Anti-christs." Nevertheless, through agreement with the city council, he abstained from changing anything in the sacramental system.

The magistracy was suspicious of Caspar Kantz's firm commitment to the Lutheran cause. (Kantz, it will be recalled, preached in Nördlingen before Billican and had met Luther in Augsburg in 1518.) Billican therefore promised to proclaim the gospel in a mild manner in order not to offend the bishop and Emperor. Nevertheless, the course of events, especially the demands made by the weavers guilds called for stronger reforms. For a short time, Billican flirted with Carlstadt's radical ideas, but renounced them in 1525, the same year he published the church order for Nördlingen. In the following years, his position toward the Zwinglian type of communion changed, not unlike that of his influential friend, Urbanus Rhegius of Augsburg.[1]

In early 1526, Zwingli and Oecolompadius began their intensive letter campaign to convince Rhegius and Billican of the rightness of their ideas. Ludwig Hetzer, Zwingli's agent in Augsburg, tried the same with words, while Luther remained idle. Billican did not fully accept Zwingli's and Oecolompadius' ideas, but preferred a middle-of-the-road position between the Wittenberg and the Zurich reformers.[2] Since Billican was neither Lutheran nor Zwinglian, he was held in contempt by both sides. Beginning in 1527, he again leaned toward the Catholic Church. When he consulted Wittenberg in 1529 about the possibility of earning a doctorate in theology at that university, Melanchthon replied: "Unless [our] university knows the religious orientation of the applicant, no degree can be conferred." Billican's wavering stand between the Catholic and Protestant faiths mirrored the city's ecclesiastical ambivalence: Communion was administered in one as well as in the other manner; vigils and mass for the dead were continued. In 1530, Billican returned to the Catholic fold, resigned from the preaching office, and became a merchant. A year later he reversed his stand and once again became Nördlingen's Protestant preacher.[3]

---

[1] Cantz, "Caspar Kantz und die Nördlinger Reformation," p. 166.

[2] Gerhard Uhlhorn, Urbanus Rhegius: Leben und ausgewählte Schriften (Elbersfeld: Friderichs, 1861), pp. 100-104.

[3] Rauscher, "Die Prädikaturen in Württemberg vor der Reformation," p. 205.

Ellwangen

Here, the preaching office at the Stift was founded in 1499 by a clergyman from Nördlingen. The occupant received an income of 100 guilders and was required to possess a doctorate in theology. Originally, it was a life-time position, but because of the unrest during the years of reformatory activity, the occupant signed a contract for either six or eight years.[1]

Johann Kress, the local preacher since 1521, switched comparatively late to the Lutheran side, that is, in the year 1525. He had received all his academic degrees, including a doctorate in theology from Tübingen, where he also served as Rector of the university.[2] Whether Melanchthon, to whom he conferred the M.A. in December 1514 had cast any influence on Kress is highly doubtful.[3] More significant was the influence of a colleague, Georg Mumpach. Mumpach was enrolled at the university in Heidelberg with Kress in March of 1514,[4] but how long he stayed is unknown. At any rate, either informal contact with Luther, or direct contact with Brenz and Billican certainly must have taken place. At the time Kress took office, Mumpach served as parson at the Ellwangen parish church.

Both Kress and Mumpach were put to death for their Lutheran stand. Kress the preacher renounced his faith and was buried in consecrated soil, whereas Mumpach refused to accept the sacrament under one form and was therefore buried "in the field."[5]

Heilbronn

This imperial city dominated the northern or lower Württemberg territory (Unterland). It was as large as Schwäbisch-Hall further to the east, but was subordinate as a Reformation diffusion center. A bridgetown across

---

[1] Josef Zeller, "Aus dem ersten Jahrhundert der gefürsteten Propstei Ellwangen (1460-1560)," Württembergisches Vierteljahrheft für Landesgeschichte, N.F., XVII (1908), 278, 280.

[2] Kuhn, Die Studenten der Universität Tübingen zwischen 1477 und 1534, I, 193.

[3] Hermelink, Die theologische Fakultät in Tübingen vor der Reformation 1477-1534, p. 203.

[4] Toepke, Die Matrikel der Universität Heidelberg von 1386 bis 1662, I, 494.

[5] Beschreibung des Oberamts Ellwangen (Stuttgart: Kohlhammer, 1886), p. 497.

the lower Neckar, it was oriented toward Speyer and Frankfurt/Main in the northwest, and toward Nuremberg in the east.[1] It had adopted the Speyer city charter in the late thirteenth century, and served as a terminal and major Neckar port. The Neckar shippers were united in a brotherhood headquartered in Heidelberg; people from Heilbronn, Neckarsulm, and other towns along the river were members of this interest group.[2] While commercial ties with the Rhein-Main and Nuremberg regions were well developed, communication with the Swabian cities of Stuttgart and Esslingen was relatively weak. When the Duchy of Württemberg announced its intention of making the Neckar between Stuttgart and Heilbronn navigable, the imperial city strongly protested for fear that the project would weaken its role as transportation depot of the lower Neckar.

Although Heilbronn was a potential center for dispersing the new faith, it lacked necessary devoted personnel. Hence, it acted more as a receiver than a giver of the evangelical message. It was influenced by Heidelberg and the Kraichgau, as well as by Schwäbisch-Hall. While 28 Heilbronners attended the university in Heidelberg between 1515 and 1524, only six went to Tübingen, and four to Freiburg and Wittenberg, respectively.

In 1520, Johann Lachmann succeeded Dr. Johann Kroener as preacher in the local St. Kilian's Church. As the son of a famous bell-founder, Lachmann enrolled as a theology student at Heidelberg in 1510. How long he studied there is not known, but we do know that he was a member of a larger group which later was to play a vital part in the Reformation movement in various Franconian cities and towns. Among the group were Melanchthon, Erhard Schnepf of Heilbronn, Johannes Brenz of Hall, Oecolompadius of Weinsberg, and Jakob Ehinger of Ulm.[3] Lachmann's clever and determined performance convinced the undecided magistracy of the necessity to carry through the changes that finally led to the abolition of mass in December, 1531.

Yet, Lachmann was rather slow in effecting ecclesiastical changes. To a large degree this depended on reformers of neighboring cities. Bucer was

---

[1] Karl Weidner, Die Anfänge der staatlichen Wirtschaftspolitik in Württemberg, Darstellungen aus der württembergischen Geschichte, No. 21 (Stuttgart: Kohlhammer, 1931), p. 26.

[2] Karl Löffler, Geschichte des Verkehrs in Baden (Karlsruhe: Keller, 1927), p. 29.

[3] It is unfortunate that most of Lachmann's correspondence with these and other persons has either been lost or not published.

quick to complain about the slow headway the Reformation was making in Heilbronn, laying the blame on Lachmann, whom he labeled as a helpless and weak follower of Brenz, unable to meet the demands of his office.[1]

The German mass was first performed on Epiphany in 1525 in the parish church by a priest from Schwäbisch-Hall.[2] In the same year, a group of local citizens demanded the Lutheran communion in both forms, which was denied by the Catholic parish administrator. On January 7, 1528, the Heilbronn magistracy asked the city government in Schwäbisch-Hall for advice on how to initiate communion under both forms, since the latter had already altered the ceremony toward the end of 1525. The Hall magistracy expressed its delight, stating that as secular authority, it had been trying to prevent disunity and open hostility among its citizens. It therefore did not officially recommend taking the sacrament under both forms, but finally gave way to it under popular pressure. In a decree, the two town councils of Heilbronn followed suit at the end of January, 1528.

The first Protestant communion in early May, 1528 was attended by 32 men and 46 women, including the city's mayor. A week later, the number of communicants had risen to 100. Wilhelm Doel (Doll) administered the communion, as no other priest seemed willing to do it.[3] Around Corpus Christi Day of that year, the city council suddenly ordered a halt to the new communion, either because it feared reprisal by the bishop, or could not afford the financial outlay.[4]

On the recommendation of Johannes Brenz in Hall, Caspar Gräter became schoolmaster and teacher of Hebrew, Greek, and Latin in Heilbronn in 1527.[5] He gave up his position as private tutor to the knight Dietrich of Gemmingen.[6]

---

[1] von Rauch, ed., Urkundenbuch der Stadt Heilbronn, XX, 766.

[2] Beschreibung des Oberamts Heilbronn, ed. K. Statistisches Landesamt, I (Stuttgart: Kohlhammer, 1901), 99.

[3] Ibid., pp. 418-19. It should be recalled that Lachmann was not in a position to serve communion, since he was foundation preacher. His sole task in Heilbronn was holding sermons.

[4] Karl Jäger, Mitteilungen zur schwäbischen und fränkischen Reformationsgeschichte (Stuttgart: Kohlhammer, 1828), p. 107.

[5] Gräter had spent some time with Brenz prior to his appointment in Heilbronn. Theodor Kolde, "Kaspar Greters Berufung nach Ansbach," Beiträge zur bayerischen Kirchengeschichte, V (1899), 199.

[6] Julius Hartmann and Karl Jäger, Johannes Brenz (Hamburg: Perthes,

In 1528, Gräter and Lachmann published the first printed evangelical catechism. In the same year, the bishop denied Wilhelm Doel and Johann Bersig the right to collect an offering for handing out communion under both forms, and for assisting Lachmann in his efforts to bring total reform to Heilbronn.

At the Speyer Diet in 1529, Heilbronn did not join the protesting cities led by Nuremberg, Ulm, and Reutlingen. Instead, it seemed to follow neighboring Schwäbisch-Hall in its active pursuance of Luther's teachings. The two cities were not only close partners in religious matters, but also shared experience in other areas, such as politics and economics.

Following the Augsburg Diet in October of 1530, both city councils ordered the "true proclamation" of the gospel; they convened the burghers on November 24 in order to ascertain their position.[1] They fully agreed with this decision. One year later (December 8, 1531), the burghers of the four Heilbronn quarters voted in favor of abolishing mass. Nonetheless, Mayor Riesser made certain concessions in favor of preserving the traditional flavor in church rituals, such as hymns and communion. The liturgy should be performed in such a manner that it would be difficult to say whether the mass was accepted or rejected. Such a wavering position would have been unthinkable in the Zwinglian-oriented southern part of the German Southwest.

Only one month later, Martin Bucer came to Heilbronn to talk to Lachmann and other local reformers. He tried to find a solution to the annoying dispute between Luther's and Zwingli's ideas of communion which had been plaguing many cities in southwestern Germany. Why Bucer chose to visit the Heilbronn reformer only months after he had made disrespectful comments about Lachmann remains a riddle. All the same, the Kraichgau with neighboring Heilbronn seemed an ideal meeting place for such an undertaking. The Kraichgau, it will be recalled, was the gateway to the upper Rhine region and the interior Neckar Basin and Franconia. It was therefore no accident that the residence of the knights of Gemmingen became the place where the Zwinglian elements of Strasbourg and the Lutheran elements of Franconia met. Because this area had

---

1840-1842), I, 187. Gräter enrolled at Heidelberg university on June 2, 1520, received his B.A. two years later, and his M.A. four years hence. See Töpke, Die Matrikel der Universität Heidelberg von 1386 bis 1662, I, 523.

[1] Ulm and Memmingen likewise held a referendum in early November.

[2] Moriz von Rauch, "Johann Riesser: Heilbronner Reformationsbürgermeister," Historischer Verein Heilbronn, XVI (1925-28), 18.

no prestigious Lutheran preacher, the talks had a chance of not being deadlocked a priori. Progress with Brenz in this respect seemed futile, since he was the author of the Lutheran Syngramma suevicum. In this connection the only three pro-Zwinglian ministers in the Kraichgau, Johann Walz, Martin Germanus, and Andreas Weinstein, visited Ambrosius Blarer in Esslingen in 1532. They begged him to counter the "anti-Zwinglian slander" on the part of Franciskus Irenicus and a number of other local Lutheran preachers. Blarer considered it his duty to help "his brothers" defend a common cause.

Although Bucer's basic attempts in reaching a settlement in the Kraichgau failed both times, he had succeeded in winning the three preachers at the Concord talks, held at Guttenberg castle outside of Wimpfen. Walz soon moved to Ulm on Blarer's recommendation.[1] He had been a co-signer of the Syngramma suevicum. Martinus Germanus had likewise adopted the Upper German strain of Protestantism. But Bucer's talks with Lachmann brought no tangible results towards an amiable settlement, since Lachmann's attitude toward Bucer reflected Brenz' negative assessment.[2]

Weinsberg

Weinsberg was a small town occupied mostly by wine growers and nestled along an ancient road that led from Wimpfen and Heilbronn to Ansbach, Hall, Ellwangen, and beyond. Its products sold mostly to Nördlingen's Ries area, to Nuremberg, and the Upper Palatinate, as well as to the Augsburg region. The town had little wholesale business and only a few artisans; still its inhabitants had reached a considerable level of prosperity which was all but wiped out during the Peasants' uprising in 1525. For this reason the numbers in Tables 1 and 2 do not coincide with the economic wealth which this community enjoyed during the earlier decades. Weinsberg's students preferred nearby Heidelberg university over any other institution.

The preachership in this town was founded in 1511 by Johann Oecolompadius (Heusgin, Huszgen, Hussschein), the father of the renowned Basle reform-

---

[1] Letter of A. Blarer to M. Bucer on April 24, 1532. See Traugott Schiess, ed., Briefwechsel der Brüder Ambrosius und Thomas Blarer (Freiburg: Fehsenfeld, 1908), I, 341-42.

[2] Brenz gathered his colleagues in the Kraichgau and Heilbronn to Weinsberg on August 15, 1532 in order to paralyze Bucer's attempts to settle the differences. To orthodox Lutherans surrounding Brenz, the Upper German reformers Blarer and Bucer and their followers were Schwarmgeister, or emotional enthusiasts.

er. It was actually the latter who had requested the creation of the preaching office.[1] As a rule, the founder always tried to preserve his link to the foundation by reserving the right to occupy the position himself or by his children.[2]

The young Oecolompadius earned a M.A. degree in Heidelberg (1501?) before he was chosen for the preaching office in April, 1510. He interrupted his career several times in order to continue his studies in Tübingen, Heidelberg, and Basle. He resigned from the preaching office in 1520. The rumor that Oecolompadius proclaimed Lutheran sentiments in his Weinsberg sermons in 1518 seems unlikely.[3] First, he spent the summer months and part of the remaining year in Basle, where he earned a doctorate in theology. In December of that year he accepted the preachership in Augsburg. It seems clear that he felt ready to preach from the Lutheran point of view only after a two-year reflection period in a monastery near Augsburg. Not until May of 1522, did he hold his first Lutheran sermon on the Ebernburg.

Oecolompadius was succeeded in June of 1520 by Erhard Schnepf, the son of a respected patrician couple in Heilbronn. Schnepf had studied jurisprudence in Erfurt between 1509 and 1511, joining the humanist circle there (sodalidas) before continuing with theological research at Heidelberg. Luther's debate in the Augustinian monastery in Heidelberg left a deep impression on Schnepf. He was selected by the mayor and magistracy of Weinsberg to become the preacher for the town.

Schnepf preached Luther's ideas until the Austrian government forced him to leave town. However, the Reformation movement continued in spite of his removal. For example, in June of 1524, Brenz told Oecolompadius in a letter that he had just met with several friends[4] in Weinsberg, and there found a genuine interest in the new teaching. He also suggested that Oecolompadius would be welcome should he return to Weinsberg.[5]

---

[1] Rauscher, "Die Prädikaturen in Württemberg vor der Reformation," p. 197.

[2] Ibid., p. 159.

[3] Sigel, Das evangelische Württemberg, Vol. VIII, No. 2, p. 856.

[4] The future signers of the Syngramma suevicum?

[5] Sigel, Das evangelische Württemberg, Vol. VIII, No. 2, p. 858. It should be remembered that at this time Protestants still formed a united front.

Mosbach

The town of Mosbach displayed its Lutheran bias as early as 1520.[1] The resident preacher, Wendel Kretz (B.A. and M.A. from Heidelberg in 1514 and 1515), held Protestant sermons at the Julianenstift beginning in 1524 or earlier. It is now certain that Kretz occupied a preaching office which he left when Elector Ludwig V suppressed the Lutheran movement in 1532.[2] Otherwise he hardly would have become the second preacher Heilbronn hired in that year.

Although we lack definite data about the channels through which the new religion reached Mosbach, it can be surmised that the town relied on Heidelberg. This comes as no surprise since Mosbach enjoyed special trade connections, importing fruits and duty-free wine from the city lower down the river.[3] Furthermore, between 1515 and 1524, almost all of the students of Mosbach attended Heidelberg university, while none matriculated in Tübingen, Freiburg, or Wittenberg. Kretz' mutual friendship with Erhard Schnepf, who, too, studied in Heidelberg at the same time was probably of decisive importance.

Brackenheim

A small wine-growing town southwest of Heilbronn, Brackenheim was located along a major thoroughfare leading from Heidelberg and Speyer through the Kraichgau into Württemberg and Bavaria. Until 1450, it served as a transit road to Frankfurt for merchants from Upper and the interior of Swabia, but it soon lost this function to the Speyer-Bretten-Cannstatt road further west.[4] Nevertheless, the road remained the most important connection between the Kraichgau-Heilbronn region and the core area of Württemberg. Gemmingen, Bönnigheim, Bietigheim, and Asperg all were located along its edge, thus serving as channel for communication between Esslingen and Ulm on the one hand, and the

---

[1] Erich Keyser, ed., Badisches Städtebuch (Stuttgart: W. Kohlhammer, 1959), p. 126.

[2] Theobald Freudenberger, Der Würzburger Domprediger Dr. Johannes Reyss, Katholisches Leben und Kämpfen im Zeitalter der Glaubensspaltung, No. 11 (Münster: Aschendorffsche, 1954), pp. 17, 19; and Neu, Pfarrerbuch der evangelischen Kirche Badens von der Reformation bis zur Gegenwart, II, 203.

[3] Richard Schröder and Karl Köhne, eds., Oberrheinische Stadtrechte, 1. Abtlg., 4. Heft (Heidelberg: Winters, 1898), p. 586.

[4] Meinrad Schaab, "Strassen und Geleitwesen zwischen Rhein, Neckar und Schwarzwald im Mittelalter und der früheren Neuzeit," Jahrbücher für Statistik und Landeskunde von Baden-Württemberg, IV (1958), 54-75.

Kraichgau on the other. The fact that the preachership in Brackenheim was held by a person from the vicinity of Ulm, as well as the surprisingly large number of Brackenheim sons enrolled at the university in Tübingen, points to powerful influences from the south. Had the occupant preacher not been removed by the Habsburgs, the town could well have become a bastion of Zwinglianism within Lutheran sovereign territory.

The preachership at St. James Church[1] had a Lutheran occupant as early as 1520: Conrad Sam. Established by a local priest in 1512, the office was endowed with at least 100 guilders, the highest in the Duchy of Württemberg.[2] In return, it was expected that the occupant be a doctor of the Holy Scripture.[3] Here, as in many other places, the relationship between the preacher and the parson was a strained one, since neither was obligated to assist the other.[4] The founder of the preachership had transferred the patronage and the right to nominate to the mayor, court, and magistracy of the town. Thus, while the right to nominate a candidate lay exclusively in the hands of the Brackenheim burghers, the town shared the presentation right with the University of Tübingen.

Sam was enrolled at the universities of Freiburg and Tübingen[5] between 1505 and 1509. With the help of the latter and Johann Oecolompadius he obtained his position in Brackenheim in 1515. Where he had earned his licentiate is unknown.[6] As Luther joyfully indicated in his letter to Sam of October 1520, Sam was one of the first Lutheran preachers in the Duchy of Württemberg.[7]

As mentioned previously, it was Johannes Gayling of adjacent Ilsfeld who informed Martin Luther about the evangelical sermons of Sam. The Brackenheim preacher must have read Luther's writings two years before he preached them. He also must have shown great enthusiasm in discussing them with a

---

[1] Unlike in most cities and towns, the foundation in Brackenheim was not established at the parish church.

[2] Rauscher, "Die Prädikaturen in Württemberg vor der Reformation," pp. 159, 176.

[3] Ibid., p. 174.   [4] Ibid., p. 173.

[5] Mayer, Die Matrikel der Universität Freiburg i. Breisgau von 1460-1656, p. 164.

[6] Sam may have received his licentiate at the University of Heidelberg, but the enrollment list does not indicate it.

[7] Beschreibung des Oberamts Brackenheim, ed. by Königliches Statistisch-topographisches Bureau (Stuttgart: Lindemann, 1873), p. 185.

man who had been close to Luther in Wittenberg.[1] Despite the constant burden of powerful opposition, Sam preached until May, 1524, when the Austrian government, on request by the local priest or parson, forced him to leave Brackenheim. Ironically, just before that he had given shelter to Eberlin von Günzburg.

Gemmingen

Here, the knight Wolf von Gemmingen became the convert and devoted servant of Lutheran doctrine in 1521. The local preaching foundation had been set up in 1510 by his father, Pleikard von Gemmingen. The position was then occupied by Bernhard Griebler, an M.A. graduate of the university in Heidelberg in 1507.[2] The enlightening moment for Wolf von Gemmingen came when Luther took his strong stand at the Diet in Worms, an event which he had witnessed and which made him a determined supporter of the Lutheran cause. He established a Latin school in Gemmingen in the same year and appointed Griebler as schoolmaster. The school soon became the educational center and diffusion headquarters for the new faith in the Kraichgau.[3]

Wolf von Gemmingen was actively involved in the Zwinglian-Lutheran dispute about the nature of the communion as early as 1525. He corresponded with Capito and Bucer in Strasbourg[4] one month after he had contacted Brenz in Schwäbisch-Hall.[5] The Strasbourg reformers expressed the suspicion that the brothers of Gemmingen (Wolf, Dietrich, and Philipp) had tried to "annihilate Christ's Word" with human opinion. In their reply, the knights deplored the divisiveness of the issue at hand, but stressed the literal interpretation of the metaphor: "This is my body."[6]

Until 1531, the offices of the Catholic priest and the Frühmesser[7] existed side by side in Gemmingen with the Protestant preacher. However, as soon as

---

[1] Otto Conrad, "Johann Gayling von Ilsfeld, ein Reformator Württembergs," Blätter für Württembergische Kirchengeschichte, N.F., XLIII (1939), 17-18.

[2] Gustav Bossert, "Beiträge zur badisch-pfälzischen Reformationsgeschichte," Zeitschrift für die Geschichte des Oberrheins, N.F., XVII (1902), 81.

[3] Tony Fleck, Gemmingen: 769-1969 (Gemmingen, 1969), pp. 58, 60.

[4] Pressel, Anecdota Brentiana, pp. 8-24.

[5] Ibid., pp. 2-6.   [6] Ibid., pp. 8-24.

[7] Can be translated as "early mass priest"; he was a subordinate helper of the priest who only read the early morning mass during the week.

the Catholic parson decided to leave, Wolf appointed Wolfgang Buss of Gernsbach in 1530, and Franciscus Irenicus of Baden-Baden, both Lutherans. Mathäus Chyträus (Kochhaf), likewise a Protestant and parson of Ingelfingen, returned in 1530 to the Kraichgau, where he was given the benefice at Menzingen.[1]

Pforzheim

This city was by far the largest (between 3000 and 4000 inhabitants) in the Margraviate of Baden. Besides the major cities of Heidelberg and Speyer, Pforzheim served as a vital commercial center, with one important road coming in from the West and connecting the city to Ettlingen and Hagenau in Alsace (Fig. 14). As a result of being situated within such a good road system, one of the city's most successful business leaders, Heinrich Göldli cultivated close personal and commercial ties with neighboring Bretten as well as with Speyer and Heilbronn. The subsequent principal of the Latin school was to come from Hagenau, and most of Pforzheim's students attended Wittenberg and Heidelberg universities, while Freiburg attracted not so many. In short, Pforzheim seemed to be oriented primarily to the north and west.

The preachership in Pforzheim's St. Michael's Chapter was handed to Johann Unger by Margrave Philipp in 1524. Unger was from a well-known patrician family in the city, attended the local Latin school, and then studied theology.[2] The Latin school was the best of its kind in the entire Southwest[3] until it was overshadowed by the school in Strasbourg. As soon as he finished his education, Unger went to Bretten and became Philipp Melanchthon's private tutor. In 1511 he returned to Pforzheim and became principal of the Latin school. During this period Unger established ties of friendship to all the persons who played important roles in the early Reformation.[4] In 1527, Margrave Philipp even allowed Unger to be married. What is more, he was permitted to return to his preaching post, even though Philipp had him put in jail for his religious views! When Unger traded his administrative position for the preaching post, Michael Hilsbach assumed the role of principal at the Latin school. He had

---

[1]Kuhn, Die Studenten der Universität Tübingen, I, 571.

[2]At which university Unger had studied is not known.

[3]Among its students were the following: Philipp Melanchthon, Johann Schwebel, Caspar Glaser, Simon Grynäus, Berchtold Haller, Caspar Hedio, and Franciscus Irenicus.

[4]Stolz, Geschichte der Stadt Pforzheim, p. 47.

held a similar post in Hagenau, Alsace, but was forced to leave the town in 1525 because he headed a small group of hard-core Lutherans there.[1] Who offered Hilsbach the high position at the famous school is not clear. When urged to partake in the mass and the Ave Maria following a decree in June of 1531, he joined Johann Schwebel as deacon in Zweibrücken.

Weil der Stadt

To the west of Stuttgart, where Franconia cuts sharply into Alemannic territory, lay the imperial city of Weil der Stadt. The town, with its 200 burghers was known for its highly developed goldsmith profession that brought considerable wealth to the community. Goldsmiths travelled widely and hence came more in contact with the new faith than any other profession.[2] The mayor of Weil, Martin Brenz, was a goldsmith. Most Weilers sent their sons to Tübingen and Freiburg to get a higher education; only a single one attended Heidelberg, and none attended Wittenberg.

The mayor and magistracy founded the preachership in 1478, with the consent of the parson at the parish church. The preacher to be hired was introduced to the abbot of Hirsau, who presented him to the bishop of Constance. The following qualities were requisite: (1) he should preach well; (2) his sermons should be understandable to the common man;[3] (3) he should live a commendable life; and (4) he should possess an advanced academic degree. His selection proceeded through consultation with two universities: Heidelberg and, very likely, Tübingen. The preacher, moreover, was ordered to abstain from regular tasks that would hinder his sermons or cause dissent with the parson and the monks.[4]

The distinguished Württemberg reformer and scholar from Schwäbisch-

---

[1] Luzian Pfleger, "Michael Hilsbach, ein oberrheinischer Schulmann des 16. Jahrhunderts," Zeitschrift für die Geschichte des Oberrheins, N.F., XX (1905), 252-59.

[2] Hans Rösler, Geschichte und Strukturen der evangelischen Bewegung im Bistum Freiburg, 1520-1571, Einzelarbeiten aus der Kirchengeschichte Bayerns, No. 42 (Nürnberg: Verein für Bayerische Kirchengeschichte, 1966), p. 72.

[3] This regulation, also to be observed in other towns, tried to curb one common abuse in the fifteenth century, namely the scholastic hair-splitting between representatives of the nominalist and realist schools. See Rauscher, "Die Prädikaturen in Württemberg vor der Reformation," p. 155.

[4] Beschreibung des Oberamts Leonberg, 2nd ed., II (Stuttgart: Kohlhammer, 1930), 1073.

Hall, Johannes Brenz, was a member of one of the most powerful patrician families in the area. A convert as a consequence of Luther's debate in the Augustinian monastery in Heidelberg, Johannes Brenz was a potent agent in propagating the Reformation in his hometown. Moreover, his father, Martin Brenz, as the Mayor of Weil der Stadt, occupied the most important position in the city government. In this context, the Austrian sovereignty regarded Martin Brenz as being the originator and opinion leader of the Lutheran movement before the Reichskammergericht in Esslingen. The government exerted such pressure on the magistracy that Brenz and his wife were not allowed honorable burial on consecrated ground.[1]

The large crowds mustered by that town's first Protestant preacher Billican can be explained only by the support offered by leading citizens to the new ideology.[2] Billican was preceded by a Lutheran parson, Johann Diepold ("Dollfuss"), who was a native of Ulm. Diepold had studied in Tübingen and delivered evangelical sermons in Weil as early as 1520.[3] His sermons, like those he later offered in Ulm, were free of emotional attacks on the Catholic Church and were met with great approval by the parish.

In 1522, the evangelical movement was given a boost by the appointment of Theobald Billican (Gerlacher) to the office of town preacher. A very close friend of Johannes Brenz, Billican had also heard Martin Luther's debate in Heidelberg. It seems almost certain that Mayor Martin Brenz was responsible for Billican's call to Weil in 1522. His sermons, directed against mariolatry, saints, and purgatory drew large crowds and such strong support from the leading figures of the town that the Austrian government was forced to intervene. The magistracy dismissed Billican and Diepold in the same year. Nonetheless, the Reformation movement continued, although somewhat abated. Many inhabitants of Weil der Stadt flocked to Reutlingen, where they could receive communion in both forms.[4]

---

[1] Sigel, Das evangelische Württemberg, Vol. VIII, No. 2, p. 739.

[2] Gustav Bossert, "Kleine Beiträge zur Geschichte der Reformation in Württemberg," Blätter für Württembergische Kirchengeschichte, VIII (1904), 148.

[3] Kuhn, Die Studenten der Universität Tübingen, I, 211. According to a source from the year 1616, all Protestants were buried on a lot on the other side of the Würm river, until approximately 1576. See Beschreibung des Oberamts Leonberg, I, 1078, footnote.

[4] Beschreibung des Oberamts Leonberg, I, 1077.

Upper and Middle Neckar Region

This Alemannic territory included the core area of Württemberg around Stuttgart, which turned into a bastion of the new faith after the political reformation of the Duchy of Württemberg. Lacking an independent diffusion center, the people of this territory turned to such outlying centers as Strasbourg, Ulm, and Zurich-Schaffhausen. Consequently, the Zwinglian coloration which these cities acquired over the year 1524-25, meant a similar orientation for the cities and parishes in the upper and middle Neckar basin. These urban communities had adopted a different constitution from that of the Franconian imperial cities. Whereas the lower and middle classes in Alemannic cities succeeded in increasing their political leverage in the fifteenth century, the patricians in Franconian cities such as Rothenburg, Nuremberg, and Frankfurt were able to preserve and extend the old constitution into the modern period.[1]

Stuttgart

Stuttgart, the capital of the Duchy of Württemberg, inherited two preacherships. They did not, however, exist side by side until the year 1534. The preaching office at the Stiftskirche was founded in 1429, was terminated in 1510, and then was reinstituted in 1534. The year 1511 saw the establishment of another preachership, that of St. Leonhards Church,[2] which was occupied by Dr. Johann Mantel from 1511 until 1515, and again from 1520 until 1523. This particular office was endowed with 1700 guilders, yielding an annual salary of 85 guilders.

It was Mantel who first proclaimed Luther's teachings in Stuttgart during his second period of office. As a fellow member of the Augustinian order, he had studied in Tübingen, Ingolstadt, and then Wittenberg, where he earned his doctorate in theology in 1507. He taught theology with Staupitz and Luther until his appointment in Stuttgart in 1511. Having become familiar with Luther's teachings in Strasbourg,[3] Mantel opposed "Good Works" by upholding the princi-

---

[1] Ludwig Fürstenwerth, Die Verfassungsänderungen in den oberdeutschen Städten zur Zeit Karls V (Göttingen: Dieterich'schen, 1893), p. 71.

[2] Rauscher, "Die Prädikaturen in Württemberg vor der Reformation," p. 194.

[3] Nikolaus Gerbellius and Otto Brunfels, Lukas Hackfurth and Philippi von Rumsperg were early adherents of the Protestant faith in Strasbourg. See Adolf Baum, Magistrat und Reformation in Strassburg (Strassburg: Heitz und Mündel, 1887), p. 8.

ple of sola fide, or justification by faith alone. He lent support to his colleagues Michael Stiefel and Johann Lonicer in adjacent Esslingen. The Austrian authorities imprisoned Mantel in 1523 for his revolutionary stand, but he was liberated two years later by rebelling peasants.

Affinities between the evangelical Augustinian order in Stuttgart and Esslingen may have existed even prior to Mantel's arrival. His predecessor, Dr. Hieronymus Gandelfinger, a native of Esslingen, had personal ties to Stuttgart and remained there until the beginning of 1523,[1] when he moved to Ulm to be near his colleague Wolfgang Rychard. Whether it was Gandelfinger or Johann Mantel who influenced the Esslingen chaplain, Martin Fuchs, is uncertain.

Mantel's arrest and confinement in Nagold drew a sharp protest from the city of Zurich. It accused the Austrian government of acting against the order of the Nuremberg Diet that demanded the preaching of the Word of God. When Luther was informed about Mantel's fate, it was not through the people of Stuttgart, but rather through one of Esslingen's burghers, Hans Schweick. He tried everything in his power to free the preacher, including petitions addressed to Duke Ferdinand and the Imperial Government in Esslingen under Margrave Philipp; but to no avail. After Mantel's release by the peasants in late April, 1525, he went to Esslingen, where friends recommended him to Margrave Philipp.[2]

Waiblingen

Like Schorndorf and other Rems valley cities, Waiblingen prospered from the sale of wine to Upper Swabia and Bavaria; it received salt in return. The majority of Waiblingen students went to Tübingen between the years 1515 and 1524, with a few also going to Heidelberg and Freiburg. No one matriculated in Wittenberg. The reason Waiblingen did not accept the Lutheran faith can only be explained by its proximity to nearby Esslingen, which had adopted the Zwinglian creed.

Waiblingen followed nearby Schorndorf in creating a preachership in 1462. A local priest donated 800 guilders plus a certain amount of wine and grain toward its procurement. According to Rauscher, however, the position yielded only 20 guilders plus an indefinite amount of agricultural produce. The ruler of

---

[1] Gustav Bossert, "Zur Geschichte Stuttgarts in der ersten Hälfte des sechzehnten Jahrhunderts," Württembergische Jahrbücher für Statistik und Landeskunde, 1914, p. 154.

[2] Ibid., p. 155.

Württemberg and the city of Waiblingen were its patrons.[1]

It is estimated that ever since the year 1515 or 1527 at the latest, the position of preacher was occupied by Leonhard Wernher of Canstatt, a confessed follower of Luther's teachings. In April of 1525 a letter by the citizens of Waiblingen testified that the divine word of God was not being impeded in any way.[2] Wernher possessed a licentiate in theology and served as dean of the Faculty of Arts in Tübingen in 1503-1504. The chaplains in Waiblingen refused to volunteer any information about Wernher to the dean of the rural chapter without the latter's consent. Nevertheless, in September of 1528, Wernher was forced out of his position and was offered another one in Esslingen. Instead, he went to Ulm.[3] Wernher probably was influenced by the course of the Reformation in nearby Esslingen.

Reutlingen

This imperial city with somewhat less than 5000 inhabitants became the only urban center south of the Franconian-Alemannic border that remained Lutheran in its religious orientation. A look at the attendance records of various universities visited by Reutlingers helps to explain this anomaly. Whereas Tübingen, with nine students, was the single most attractive school between 1515 and 1524, Heidelberg and Wittenberg each matriculated five, while Freiburg attracted only two pupils.

The local preachership was created by the magistracy in 1518, but as in Nördlingen and other towns[4] the patronage did not belong to a person or corporation in the city, but rather was the property of Königsbronn monastery outside the city walls. Since this institution did not provide the qualified and serious-minded personnel demanded by the burghers,[5] the city established its own

---

[1] Rauscher, "Die Prädikaturen Württembergs vor der Reformation," p. 80.

[2] Sigel, Das evangelische Württemberg, Vol. VIII, No. 2, p. 527.

[3] Gustav Bossert, "Die Jurisdiktion des Bischofs von Konstanz, 1520-1529," Württembergische Vierteljahreshefte für Landesgeschichte, N.F., II (1893), 276.

[4] Notably Dinkelsbühl, Rothenburg/T., Miltenberg, and Isny.

[5] In 1514, the citizens of Reutlingen complained about the dean of the local chapter, Peter Schenk, that he had not said mass, nor held any sermons for months, and that he constantly performed pranks in the parish church. Even then the city threatened to withhold the tithe by transferring it to a person of their own choice. See Beschreibung des Oberamtes Reutlingen, 2nd ed. (Stuttgart: W. Kohlhammer, 1893), II, 100.

preaching foundation. Their dependence on (and the atrocious performance by) the monastery of Königsbronn prompted many burghers of Reutlingen to adopt Protestantism.

In 1519 or 1520, the magistracy appointed the 24-year-old Mathäus Alber to the preachership. The son of a former goldsmith, Alber attended preparatory schools in Schwäbisch-Hall, Rothenburg/Tauber, and Strasbourg before enrolling at the university of Tübingen, where he became a student of Melanchthon and assistant to Brassicanus, the humanist. Together with Franciscus Irenicus, Ambrosius Blarer, and Johann Oecolompadius, Alber formed a close-knit group[1] of friends led by Melanchton the young Greek teacher and future assistant to Martin Luther. Like the Engelbrecht group in Freiburg, these humanists must have read Luther's first writings with special curiosity.

After Melanchthon's promotion to full professor in Wittenberg, in August 1518, Alber transferred to Freiburg, where he finished his baccalaureus sententarius on August 1, 1521.[2] Contact with Philipp Engelbrecht must have further strengthened Alber's conviction about the rightness of Luther's ideas. Alber also had connections to Jakob Windner, a Reutlingen resident who moved to Constance, where he began preaching the new faith in 1519. These relationships with Melanchthon and Irenicus on the one hand, and Ambrosius Blarer, Oecolompadius, and Engelbrecht on the other, largely explains the central position Reutlingen occupied in the Reformation movement.

In 1521, Alber again took charge of the preachership in his hometown Reutlingen, stressing the biblical message in his sermons from the start. His bible scholarship is demonstrated by the fact that he instructed many of the Reutlingen chaplains in the Gospel of St. Matthew and the Letter to the Romans. On March 1523, Alber received an encouraging letter from Zwingli, in which the latter praised him as an outspoken and independent reformer. Zwingli ended his letter with the following plea, characteristic of his pragmatic bent: "Recommend me to your faithful brothers!" The Swiss reformer must have written this letter in the hope of winning the imperial city for the Swiss Reformation. The bearer of the letter, former Franciscan monk Konrad Hermann,[3]

---

[1] Johann Butzbach, the Protestant parson, also must have been one of the members.

[2] Mayer, Die Matrikel der Universität Freiburg im Breisgau von 1460-1656, I, 250-51.

[3] A Villingen resident and graduate of Freiburg university in 1515. See Mayer, I, 211.

was Zwingli-oriented and tried to induce Alber in that direction; but to no avail. He wanted to have a debate with him, which Zwingli sought to prevent, Alber again refusing to follow Zwingli's course. In 1525, Hermann was forced to leave Reutlingen for Ulm, where he acted as mediator between that city and Zurich.[1]

Nevertheless, Alber tried to prevent a complete break with Zwingli. Owing to the geographical location of Reutlingen between the Franconian centers of Heidelberg and Schwäbisch-Hall and the Zwinglian centers of Zurich, Constance, and Strasbourg, Alber realized that he was acting as mediator between the two opposing doctrines. This is one of the reasons why the Reutlingers and their leaders were thought to be "by far the mildest Lutherans."[2] At least in the formative years of Lutheranism, Alber showed a certain predilection for the Zwinglian communion, and even during the later years never formally embraced orthodox Lutheranism.[3]

During the fall of 1523, the Austrian authorities tried to stall the progress and diffusion of the Reformation in Reutlingen and its umland by ordering a blockade of the imperial city. Subjects of the Duchy of Württemberg were threatened with prohibition to enter and trade with the city due to the "seductive, suspicious, annoying, disagreeable, seditious, and heretical Lutheran teaching . . . over the pulpit and through other public channels."[4] Nevertheless, the magistracy continued its active support of Alber, inviting the government to disprove their leader's ideas on the basis of the Holy Scripture. On Easter, 1524, Alber abandoned the Latin mass in favor of the German mass, and held communion according to the New Testament by administering bread and wine. He married a woman from Reutlingen one month later, thus becoming the first clergyman in Württemberg to take this bold step.

Toward the end of 1525, Reutlingen sent a delegate to Luther in Wittenberg in order to gather some advice on its new ecclesiastical organization. In two letters dated January 4, 1526, Luther expressed his pleasure about the par-

---

[1] Walther Köhler, Zwingli und Luther, I, 203.

[2] In a letter of March 1532 to his brother Thomas, Ambrosius Blarer certified that the Reutlingen preachers surrounding Alber were friendlier than other Lutherans further north; because of this he tried to prevent any further alienation. See Schiess, ed., Briefwechsel der Brüder Ambrosius und Thomas Blarer, I, 334.

[3] Beschreibung des Oberamts Reutlingen, II, 114.

[4] Hans-Jörg Reiff, Reformation und Verfassung in Reutlingen, Zulassungsarbeit Tübingen, 1970, p. 18.

ish's virtuous conduct as well as changes in the ritual. Yet, he warned against too swift of a change, i.e., reverting to the Zwinglian communion, since this would anger the unstable faction within the congregation. He hurled charges against the sectarians, mobs (Rotten), heretics, and false spiritualists (Geister), primarily Karlstadt, Zwingli, and Oecolompadius. He ended the letters with a plea to the citizens of Reutlingen to remain loyal to his form of communion.[1]

The special significance of Luther's letter consists in the fact that the Wittenberg reformer gave his stamp of approval to the plain church service in Reutlingen, which had as its basis the medieval Prädikatur church service.[2] Its liturgy differed from the more complex service held in other Lutheran churches in the Southwest which had adopted Luther's German mass. This plain communal ritual without altar lights and chasubles served as the prototype for Erhard Schnepf's communion which he instituted throughout Württemberg in 1536.[3]

Göppingen

Situated directly on the Fils, a Neckar tributary, this town also profited from the wine trade along the main thoroughfare between Speyer-Esslingen and Augsburg-Munich. The manufacture and sale of cloth likewise brought prosperity to many Göppingen residents.[4] The preachership, created in 1514, was located in the parish church of Oberhofen, outside of Göppingen. It was incorporated into the provost's office and the chapter of the Oberhofenstift. The patronage lay with Duke Ulrich. Although the preacher was asked to have either a doctorate, licentiate, or at least a bachelor of the Holy Scripture, the first occupants of the office had only an M.A.[5]

---

[1] Beschreibung des Oberamts Reutlingen, II, 108, footn.

[2] Heinrich Hermelink, "Das Luthertum der württembergischen und der bayerischen Landeskirche," in Auf dem Grunde der Apostel und Propheten: Festgabe für Landesbischof D. Theophil Wurm zum 80. Geburtstag (Stuttgart: Quell, 1948), p. 152.

[3] Paul Schwarz, "Die Reformation in einer Reichsstadt: Am Beispiel Reutlingen dargestellt," Süddeutscher Rundfunk radio broadcast, October 27, 1972. Theophil Wurm, "Der lutherische Grundcharakter der württembergischen Landeskirche," Blätter für Württembergische Kirchengeschichte, Sonderheft VI (1938), pp. 23-24.

[4] Keyser, ed., Württembergisches Städtebuch, p. 100.

[5] Rauscher, "Die Prädikaturen in Württemberg vor der Reformation," p. 184.

Martin Cless held the office in Göppingen after 1516. The son of a rich burgher and customs official, he received his education in Freiburg and Tübingen. From the latter he received his M.A. in 1513. After preaching in Sindringen, Rottenburg, and Leonberg, he returned to his hometown of Göppingen. It has been said that Cless turned Lutheran during his service as parson in Leonberg between 1521 and 1524, when he had to flee the town.[1] It is difficult to establish a date for Göppingen before 1529; Cless probably proclaimed the Gospel as early as 1524 upon his arrival from Leonberg. He might have preached in a more subdued manner until the late twenties, for when his provost demanded in 1529 that he preach in the traditional way, he openly broke his ties with the old church. He fled with his mother to adjacent Ramsberg castle, where he was protected by the Knight Philipp von Rechberg.[2]

Ehingen

In this community, the preaching office was set up by two burghers from Rottenburg in 1451. In 1521(?), Johann Eycher, son of a wealthy Rottenburg family took over as its preacher. His passionate yet intelligent behavior made him an influential figure in the new movement. The Austrian government feared his influence on the local ecclesiastics and the population. His home served as the meeting place for all friends of the local reform movement. In 1523, Eycher went to Wittenberg and was succeeded by the prebendary, Conrad Wachendorfer.

Eycher probably returned to Rottenburg in 1524, since the Austrian authorities tried to curb his activities in that year. Again, in the spring of 1527, the Hohenberg government in Rottenburg informed the government in Innsbruck about his sermons and "awkward actions." In April and June, 1527, he was prohibited from preaching. He thereupon left Ehingen-Rottenburg, exchanging his position with that of Caspar Wölflin of Wannweil near Reutlingen.[3] The two Protestant prebendaries, Wachendorfer and Schuhmacher, were ordered to be jailed by Wölflin, then expelled. With this drastic countermeasure, the Austrian government set an end to all Reformation attempts in the area.

---

[1] Sigel, Das evangelische Württemberg, X, 563.

[2] Theodor Schön, "Meister Martin (Cless) von Uhingen als Prediger in Rottenburg," Reutlinger Geschichtsblätter, XIII (1902), 30.

[3] Beschreibung des Oberamts Rottenburg (Stuttgart: Kohlhammer, 1900), pp. 82-83.

Rottweil

This imperial city, with its considerable territory was the largest in the upper Neckar area. Most of its students attended nearby Tübingen (8) and Freiburg (6). Only two went to Wittenberg, while none were enrolled in Heidelberg. The affiliation with Freiburg alone suggests a Zwinglian bent for Rottweil's Protestant community. However, the strong political, economic, and intellectual ties with Switzerland were the primary factors that contributed toward its religious outlook.

In spring or summer of 1524, Dr. Anshelm, a Rottweiler, was impelled to leave Berne because of his illegal agitation on behalf of the new teaching.[1] He returned to Rottweil, where he became the opinion leader of a small group of Protestants. But the movement did not gain momentum until 1526, when Conrad Stücklin[2] of neighboring Sigmaringen came to Rottweil and in the following year delivered his first Protestant sermon. The city's guilds showed particular enthusiasm for the new ideology, but as soon as the town council became aware of the growing movement, it intervened to prevent such an innovation. A Rottweil clergyman in Constance in a letter encouraged the council to adopt the new religion.

Generally speaking, the Reformation came to Rottweil from the outside and was kept alive to some degree by persons who resided either in Constance, Berne, or Zurich. As already mentioned, an important factor was the close political ties that this city enjoyed with Switzerland.[3] In 1519, Rottweil signed an eternal political alliance with the thirteen Swiss prefectures, which was the first step toward complete political integration.

Although Rottweil did not boast of any trading companies like its neighbor Horb, it enjoyed an active trade, exporting grain to Switzerland. Weavers and

---

[1] Anshelm studied at the universities in Cracow and Lyon, where he finished his doctorate in medicine. He became schoolmaster and town-physician in Berne, all the while admiring Huldrych Zwingli. See Johann Speh, "Beiträge zur Reformationsgeschichte des oberen Neckargebietes: Rottweil und Hohenberg" (unpublished Ph.D. dissertation, University of Tübingen, 1920), p. 20.

[2] Stücklin earned his B.A. in Heidelberg in 1495, becoming chaplain in Sigmaringen in 1519. He went to Constance in 1529.

[3] Paul Kläui, "Rottweil und die Eidgenossenschaft," Zeitschrift für Württembergische Landesgeschichte, XVIII (1959), 1-14; Wolfgang Vater, "Die Beziehungen von Rottweil zur Schweizerischen Eidgenossenschaft im 16. Jahrhundert," in 450 Jahre Ewiger Bund: Festschrift zum 450. Jahrestag des Abschlusses des Ewigen Bundes zwischen den XIII Orten der Schweizerischen Eidgenossenschaft und dem zugewandten Ort Rottweil (Rottweil: Stadtarchiv, 1969), pp. 26-64.

blacksmiths sold their red cloth and their sickles in the annual fair at Zurzach, which was easily accessible via the "Swiss Road" that connected Stuttgart, Balingen, and Rottweil with Schaffhausen and Constance. It was not surprising, therefore, that the blacksmiths and weavers under their guild masters Hans Bock and Michael Furtenbach pressed particularly hard for religious change.[1] The strongest opponents of the Reformation were the patricians and the peasants of the rural environs of the city.[2]

One reason for Rottweil's strict adherence to the Catholic cause was the threat of Emperor Charles V to deprive the city of the imperial court, which brought prestige as well as revenue to the town. In August 1529, over 400 Protestants (mostly blacksmiths, clothmakers, and members of the intelligentsia) were expelled from Rottweil to settle in places like Strasbourg, Gengenbach, Constance, Diessenhofen, and other towns.[3] This caused a sharp decline in population, revenues, and business activities, not to speak of the Reformation movement there.

Upper Rhine Area

The Rhine valley has been considered the "axis" and backbone of the Holy Roman Empire, and with justice. Owing to its extensive river and road communications, it was the main traffic artery within Germany since Roman times. It formed the connecting link between the upper and lower parts of Germany as well as between the Northwest and South. Frankfurt/Main was the primary focus of the Rhine valley and mediator between these regions. Its central position within the four borders of Germany, together with its commercial and manufacturing prowess, gave it the title "capital" and "shopping mart" of the Empire.[4]

---

[1] Wolfgang Vater, "Bauernkrieg, Reformation und Ansätze zur katholischen Reformation in der Reichsstadt Rottweil" (Zulassungsarbeit, University of Freiburg/Br., 1964), p. 18.

[2] Lecture by Dr. Wolf Narr, University of Tübingen, May 12, 1972.

[3] Listed in Heinrich Ruckgaber, Geschichte der Frei- und Reichsstadt Rottweil (Rottweil: Englerth, 1894), pp. 244-45.

[4] Otto Maull, "Der Rhein-Mainische Lebensraum," in Rhein-Mainischer Atlas, ed. by Hans Behrmann and O. Maull (Frankfurt: Lehmann, 1929), p. 8.

Offenburg

This imperial city on the east side of the Rhine opposite Strasbourg functioned as a crossroad linking Frankfurt with Basle and Strasbourg with Villingen/Rottweil. However, its location did not give the town the advantage of developing into a trading center. It was too close to Strasbourg, which prevented economic expansion until the nineteenth century.[1] Although twice as many Offenburgers attended Heidelberg than Freiburg university, the impact Strasbourg had on the town made itself felt in its Zwingli-Bucerian religious outlook.

In 1525, the Offenburg magistracy appointed first one, then another preacher to the Liebfrauenkirche outside of town. Both of these men were given equal status with the parson. The Count Wilhelm of Fürstenberg was the provincial governor of the Ortenau and replaced an able priest who had tended the benefice for several years with a foreign clergyman. According to local Catholic authority, the new foundation preacher completely allied himself with the Lutheran sect and "dared to preach the evil, poisonous Lutheran ideas."[2] On the other hand, the still sizable number of "real Christians" or Catholic sympathizers of the patriciate clearly viewed it as a threat to the existing socio-political order.[3]

Nevertheless, the Reformation was solidly backed by many of Offenburg's citizens. In 1530, their representatives at the Diet in Augsburg voted for the new faith, together with neighboring Strasbourg. At the same time it declared its obedience to the Emperor. But soon thereafter, according to Gothein in 1531, a change took place: The more Strasbourg leaned toward the Protestant faith, the more the Catholic religion gained ground in Offenburg.[4] The evangelical sermons were abandoned, and the preachers left the town.

What caused this religious reversal is unknown. It can be assumed, however, that external political factors played an important role. Leading town figures speculated on increasing their wealth and power by luring Catholics from

---

[1] Keyser, ed., Badisches Städtebuch, p. 334.

[2] The designation "Lutheran" also included the Zwinglian-Bucerian brand of Protestantism. Like all the Protestant clergy in Strasbourg and the Kinzig valley, the Offenburg preacher belonged to the latter.

[3] Ernst Batzer, "Neues über die Reformation in der Ortenau," Zeitschrift für die Geschichte des Oberrheins, N. F., XXXIX (1926), 73-74.

[4] Eberhard Gothein, Wirtschaftsgeschichte des Schwarzwaldes und der angrenzenden Landschaften (Strassburg: Truebner, 1892), I, 270.

Strasbourg into town and then upgrading its prestige by becoming a gathering place for the Catholic party, right next to a large Protestant metropolis. Many constituents of the cathedral chapter in Strasbourg sought refuge in Offenburg. In spite of this, many Offenburgers were unwilling to turn back to Catholicism again. They were rather attracted to the Protestant sermons in Strasbourg; to prevent this, the magistracy ordered the city gates closed on Sundays and holidays.[1] This, however, did not abolish the preachership, whose patronage belonged to the town. Instead, it ordered a Catholic into the position, since it had come to realize that the Reformation could be countered best with its own mechanism of success: the sermon.[2]

Baden-Baden

The capital of the Baden margraviate, Baden-Baden served as a spa as early as Luther's time. It attracted many people who sought treatment of gout, rheumatism, and catarrh.[3] Students from Baden-Baden attended the universities in Heidelberg, Freiburg, and Tübingen; hardly anyone went to Wittenberg, a place little known in the Upper Rhine Plain. That the number of students was almost evenly divided between Heidelberg and Freiburg is evidence of the strong contention between the two Protestant factions.

Margrave Philipp appointed Franciscus Irenicus to the courtly preaching office of Baden in 1524. He presided until March 1531.[4] As a native of Ettlingen, he had studied and taught in Tübingen and then Heidelberg, where he became a faithful follower of Luther in 1518. Encouraged by the latter to study theology, Irenicus soon became the opinion leader of the Lutheran clergy in Baden. He was also a classmate of Vehus the court chancellor, who suggested Irenicus as court preacher to Philipp. Irenicus spent several years in Esslingen as itinerant preacher and while there, married a local burgher's daughter in 1525.[5]

---

[1] Ibid., p. 75.   [2] Ibid., p. 271.

[3] Keyser, ed., Badisches Städtebuch, p. 11.

[4] At least since 1524; that he preached the new religion in his hometown, as Vierordt asserts, is doubtful. See Gerhard Kattermann, Die Kirchenpolitik des Markgrafen Philipp I. von Baden (1515-1533), Veröffentlichungen des Vereins für Kirchengeschichte in der evangelischen Landeskirche Badens, No. 11 (Karlsruhe, 1936), p. 20, footn. 66. In 1531, Irenicus tried to find a preaching position in Esslingen, but Bucer and Blarer rejected him because of his "lutherizat de eucharista."

[5] Kattermann, Die Kirchenpolitik des Markgrafen Philipp I., p. 25.

In 1526, Dr. Jakob Strauss came to the town of Baden at the request of the palace superintendent, or the Landhofmeister. Despite his Upper German background and his evangelical activity in Tyrol, he became and remained a staunch Lutheran. Strauss was an ardent enemy of Zwingli, Oecolompadius, and the Strasbourg reformers. As early as spring of 1526, he had convinced Philipp that Oecolompadius's ideas were highly heretical. In the following months, he obtained a prohibition on all writings of Zwingli, Oecolompadius, Bucer, Hedio, as well as the rest of the upper German religious reformers.[1]

As canon and parson of the prebendary (Stiftspfarrer), Strauss made his congregation pray that God should not tolerate the denial of Christ's presence in the bread of the altar. His pamphlet, entitled "Against Master Huldrych Zwingli's severe error," appeared shortly afterwards and was a denunciation of the ideas advanced by Zwingli and the Strasbourg reformer.[2] Needless to say, it greatly annoyed the latter. Oecolompadius urged Zwingli to retort to the writer. Zwingli's reply "About Doctor Strauss' Booklet" was printed at the end of 1526. In January of the following year he added an accompanying letter addressed to Margrave Philipp. The Strasbourg reformers assumed the task of relaying both of these messages to Baden-Baden.[3] Again one year later, Strauss published a new pamphlet against Oecolompadius' Antisyngramma. Philipp likewise seemed to be opposed to the Swiss interpretation of the communion.

Indeed, in the summer of 1528, imperial pressure caused the margrave to reintroduce a number of traditional Catholic ceremonies such as the celebration of Corpus Christi Day and the exhibition of the holy grave. This ordinance led to the resignation of twenty Protestant preachers, among them Melchior Ambach, Martin Fuchs, Johann Mantel, Martin Ofner, and Balthasar Hirt.[4] That these preachers were strict followers of Zwingli became evident by their later activity in towns south of the Lutheran-Zwinglian demarcation line.[5]

---

[1] Ibid., p. 68.

[2] Vierordt, Geschichte der Reformation im Grossherzogtum Baden, p. 247.

[3] As late as 1531, the Strasbourgers referred to their choice of Protestantism "auf Zwinglische Weis' oder wie man es nennen wolle." "Die Reformation der Reichsstadt Biberach," Historisch-Politische Blätter für das katholische Deutschland, LVIII (1866), 724.

[4] Kattermann, Die Kirchenpolitik des Markgrafen Philipp I., p. 70; Vierordt, Die Geschichte der Reformation im Grossherzogtum Baden, I, 250.

[5] Mantel, while in Iffezheim, got to know Matthäus Zell and his family in the nearby spa of Baden-Baden, where the Strasbourg preacher took the baths

Their partial replacement by other Protestant preachers[1] confirms the fact that Philipp's policy was not so much directed against the new faith as a whole as against the Zwinglian elements in his territory.[2] The neighborly feeling that had developed between the margrave and the city of Strasbourg fell victim to this Zwinglian-Lutheran quarrel.

Lahr

Lahr, which was under the joint rule of Lahr-Mahlberg and Margrave Philipp of Baden was thus only in part governed by the ruler of Baden. The preachership in Lahr was established in 1497 by the shoemaker and tanners guilds for the spiritual welfare of their members and all of the town's citizens.[3] Many of their decrees were similar to the ones issued in Philipp's realm, for instance, allowing the marriage of priests and the administering of communion in both forms to dying persons. In 1525, as a result of the Peasants' Rebellion, the rulers ordered that all sermons must agree with the Bible. Between 1515 and 1524, two Lahr citizens attended Freiburg university. None went to Heidelberg or Wittenberg. The location of Lahr almost halfway on the key route between Strasbourg and Freiburg was important for its religious outlook. The foundation preacher was Jakob Ofner, a graduate of Freiburg. He enrolled there in December of 1502, only fourteen days prior to Ludwig Oeler.[4] His friendship with the latter was what might have sparked Ofner's conversion. Oeler, who was influenced by Philipp Engelbrecht, became chaplain at Freiburg

---

in 1528. Zell invited him to Strasbourg, from where Mantel went to Hesse. However, his strong Zwinglian stance was not tolerated here. He turned south again and became minister in Elgg near Zurich. Vierordt, I, 250.

[1] Among them Mornhinweg, Zachmann, Anastasius Meyer, and Balthazar Hirt (Sarhirt).

[2] Mantel's replacement by Mornhinweg in Iffezheim presents an interesting case. Mornhinweg, devout Zwinglian, must have returned to a more moderate position concerning the communion, which he expressed three years later on his journey to Esslingen. Bucer still praised Mornhinweg's piety, but also voiced his objection about the latter's change of mind. He was received in Esslingen with great mistrust. See Gustav Bossert, "Georg Mornhinweg, der erste Prediger des Evangeliums in Klosterreichenbach," Blätter für Württembergische Kirchengeschichte, N. F., XIII (1909), 132.

[3] F. I. Mone, "Predigerpfründe im 14. and 15. Jahrhundert zu Heidelberg, Lahr und Basel," Zeitschrift für die Geschichte des Oberrheins, XVIII (1865), 10.

[4] Mayer, Die Matrikel der Universität Freiburg von 1460 bis 1656, I, 148.

cathedral. He fled to Strasbourg in 1523, when the magistracy threatened to prosecute him for his Lutheran sermons. This made the Reformation movement in Lahr dependent primarily upon Strasbourg.

The parson of Lahr, Martinus N.[1] likewise was a convert. In the spring of 1528, it was decided that both Ofner and Martinus N. should be dismissed, since they preached in favor of the communion sub utraque specie despite several warnings from the town's territorial rulers. Neither Ofner's petition, nor the efforts of the Lahr burghers to save them could prevent the suppression of the Reformation movement in Lahr.[2]

Heidelberg

In 1391, the domestics of the court of Count Ruprecht II established a preachership in the Holy Ghost Church in Heidelberg. The applicant was required to hold a bachelor's degree in the Holy Scripture and was offered a salary of 100 guilders in 1412.[3] He was assigned to give sermons in German, since enough Latin sermons were being offered throughout the city and the university already.

As mentioned previously, Luther's debate at the convention of the Augustine monks in Heidelberg in 1518 won many of the younger humanists and clergy over to his side. They, in turn, helped to diffuse his teachings over southwestern Germany by converting other persons associated with the university in Heidelberg, and then migrating to other cities. The position of the counts of the Palatinate toward the Lutheran ideology was determined by political, rather than by religious considerations. It was the fear of potential political unrest that served as guide for their actions. While Elector Ludwig forced Brenz and Billican to leave Heidelberg in 1522, he allowed his court preacher, Wenzeslaus Strauss, to hold sermons at the Holy Ghost Church the following year. According to the Knight Hans of Landschad, the elector favored Luther's ideas until the Worms Diet in 1521, then withdrew his support. Only after Sickingen's defeat in 1523, when the imperial government failed to honor the elector's demands on the knight's possessions, did Ludwig endorse the demands of stu-

---

[1] In the sixteenth century, N stood for X or an unknown surname.

[2] Kattermann, Die Kirchenpolitik des Markgrafen Philipp I., p. 80.

[3] Mone, "Predigerpfründe im 14. und 15. Jahrhundert," p. 6. The brotherhood of the domestics of the court provided 40 guilders, which was supplemented with 60 guilders from the Count Palatinate.

dents, by appointing the Lutheran professors Simon Grynäus, Herman von dem Busche, and Sebastian Münster to the university. The new ideology was rapidly gaining acceptance so that Planitz could assure the Saxon elector in July of 1524: "The gospel is proclaimed purely and sincerely, but they reject the label 'Lutheran.'"[1] This rejection stemmed from no other than Luther himself, who credited the religious movement to divine inspiration.

Since Heidelberg was located at the edge of the Upper Rhine Plain, it soon began to feel the influence of the Zwinglian movement that thrust northward from Basle and Strasbourg. Some Heidelbergers, Grynäus among them, soon endorsed the Zwinglian-Bucerian mode of communion. As Grynäus wrote to Capito in April, 1526, "In Heidelberg, too, there are people who do not expect meat at the communion."[2]

Wenzeslaus Strauss received his M.A. in 1514 and was appointed dean of the faculty of arts in 1519. He kept his position until May, 1526. He personally witnessed Luther's appearance in Heidelberg. The devil-may-care attitude of the elector enabled Strauss to hold Lutheran sermons in 1523.[3] His sharp attacks against Catholicism earned him the title "our evangelical trumpet" (tuba evangelica)[4] and resulted in violent clashes with advocates of the old faith. As expected, they failed to win the support of the elector. Withstanding their all-out attempts to oust Strauss from his position, the court preacher was able to hold on to it until May, 1526.

Main and East Franconian Cities

Most of this area was ecclesiastical territory ruled directly by the prince-bishops in Würzburg, Mainz, and Bamberg. Hence most of the Main valley was called des Reiches Pfaffengasse or "ecclesiastical thoroughfare of the Empire." This explains the lack of significant local centers of spread for Luther's teaching. Even Nuremberg failed to play the role assigned to it by its sheer size, as well as its economic and personal resources. To most of Franconia this metropolis was marginally situated.

---

[1] Vierordt, Geschichte der Reformation im Grossherzogtum Baden, I, 151.

[2] Ibid., p. 235.

[3] Walter Müller, Die Stellung der Kurpfalz zur lutherischen Bewegung von 1517 bis 1525, Heidelberger Abhandlungen zur mittleren und neueren Geschichte, No. 60 (Heidelberg: Winter, 1937), pp. 127-29.

[4] Alexander Märklin used the expression in a letter to Johann Schwebel. See Bossert, "Beiträge zur badisch-pfälzischen Reformationsgeschichte," p. 53.

Rothenburg/Tauber

Here, the preachership was founded in 1468 by two local clergymen, whose aim was to improve the clerical care of the citizens. Like Reutlingen, Nördlingen, and Miltenberg, Rothenburg had a parish church, but not its own parson or foundation preacher. St. James parish church, including its revenues, were in the possession of the Teutonic Order, and was therefore obliged to provide ecclesiastical service to Rothenburg's citizens. Yet as in the case of Reutlingen and Rothenburg, the Order did not fulfill its obligations. Consequently, the city's inhabitants constantly complained about the immoral conduct of its members.[1] The new preachership was thereupon bestowed on a secular priest, who received an annual allowance of 100 guilders. The demands were that he have a doctorate or licentiate in the Holy Scripture, or experience in the art of preaching.[2]

At the time of Luther's public appearances, Dr. Johannes Teuschlein occupied the preaching office in Rothenburg. He had studied in Leipzig, was appointed professor of theology in Wittenberg, and received a doctorate in theology in 1508. But unlike Karlstadt, who arrived there in 1524, Teuschlein had no specific ties to the Wittenberg reformers at that period or later.[3] Luther's 95 Theses apparently did not move him; nor was this surprising, since he was known for his vast intellectual capacity and close friendships with equally endowed persons. An important influence not to be ruled out was Johannes Hornburg, a well-known patrician and enthusiastic follower of humanism, who studied in Leipzig and befriended a number of Lutherans, Althamer among them.[4] In 1520, Hornburg went to Wittenberg, and in April, 1521, he enrolled at the university.[5]

In 1521, the head of the Teutonic Order complained to the magistracy

---

[1] Theodor Kolde, D. Johannes Teuschlein und der erste Reformationsversuch in Rothenburg o.d.T. (Erlangen: Böhme, 1901), pp. 3-5.

[2] Helmut Weigel, Die Deutschordenskomturei in Rothenburg o.T. im Mittelalter, Quellen und Forschungen zur bayerischen Kirchengeschichte, No. 6 (Leipzig: Deichert, 1921), p. 101.

[3] Kolde, D. Johannes Teuschlein, pp. 7-8.

[4] Theodor Kolde, "Zur Reformationsgeschichte von Rothenburg/T.," Beiträge zur bayerischen Kirchengeschichte, III (1897), 174.

[5] Carolus E. Foerstemann, Album Academiae Vitebergensis 1502-1560 (Leipzig: Tauchnitz, 1841), p. 102.

about Teuschlein's German sermons. They were contrary to established practices in the bishopric and were thought to result in a schism. It was a clear sign of change in Teuschlein's thinking, which finally led to an unbridgeable gulf between the parson and the preacher. Inevitably, Teuschlein was to be accused of trying to abolish the mass.

By October, 1524, Teuschlein had experienced such a large increase in his parish that the magistracy did not dare to get rid of him. Valentin Ickelsamer[1] and Diepold Peringer,[2] both of them peasants and preachers at the Franciscan monastery, became Teuschlein's assistants. The majority of the parish, about 300 burghers, expressed their support for Teuschlein, and the influential former mayor, Ehrnfried Kumpff, also sided with the preacher.

In late fall, 1524, Andreas Karlstadt arrived in Rothenburg, where he got a friendly reception by Teuschlein and his followers. Karlstadt's lodgings served as a meeting place for their secret religious discussions. Meanwhile, Margrave Casimir of Ansbach issued a search warrant and asked the city council to expel Karlstadt. But Kumpff opposed Casimir's plans and kept Karlstadt in hiding.

The Würzburg prince-bishopric then thought it necessary to stop the galloping Reformation movement in Rothenburg by excommunicating Teuschlein, but the preacher pointed to his duty to preach the gospel and demanded that the bishop should furnish proof of his "unchristian actions."[3]

The socio-religious conflict in the city became acute when the Peasants' War broke out there in March, 1525. Led by a knight, the burghers elected a committee that soon took over governmental powers. It was mostly Karlstadt who had quietly instigated the social revolution. The magistracy was unable to cope with the widespread unrest in the city and countryside, and Kumpff expelled the Catholic priest and put an end to the mass. Violent outside interference soon brought the movement to an end.[4]

In early June, the city submitted to the forces of the Swabian Federation

---

[1] A learned Latinus who had studied in Erfurt.

[2] He also claimed to be a peasant and was involved in their rebellion. Following its suppression, Peringer was drowned. See Ernst Enders, Dr. Martin Luther's Briefwechsel, V (Calw, 1893), 154.

[3] Kolde, Dr. Johannes Teuschlein, p. 28.

[4] Ibid., pp. 32-33.

and reinstated the traditional mass on the request of 200 burghers.[1] Karlstadt and Kumpff fled when the old magistracy regained its power. Georg Vogler of Ansbach tried without success to persuade the magistracy to accept a "true evangelical preacher."

Würzburg

Würzburg, the largest city on the middle Main was situated on one of the main roads that connected southern Germany with Saxony. Like the majority of Franconian cities, Würzburg sent most of its sons to universities in Saxony and Thuringia. Between 1515 and 1525, a few went to Heidelberg (4), and Freiburg (7), none to Tübingen; at the same time, thirteen went to Wittenberg alone, not to mention Leipzig, Erfurt, and other universities.

Paul Speratus held the preachership in the local cathedral since July of 1520. The basic salary paid by its patron, the cathedral chapter and the bishop, was 80 guilders. On request of ecclesiastical circles, Speratus came from Dinkelsbühl, where he had held a similar position.[2] Nothing there indicated that he had switched from the old faith to the new. In June of 1521, the first accusations were being levelled at Speratus; in September, the number of complaints about his sermons mounted.[3] He was forced to resign in November and shortly afterwards appeared in Vienna.

Where and how Speratus was influenced in the new faith is not definitely known. One factor, but not the only one which made him receptive toward the Lutheran religion was his far-reaching travel and educational experience. In addition, his education brought him in contact with Hedio and Capito. By then, both had definitely accepted the new ideology. But whereas both of them soon adopted the Zwinglian Reformation, Speratus never left the Lutheran "party." In February, 1522, Johann Poliander was hired from Wittenberg as foundation preacher <u>ad Probam</u> (on trial). He had mutual connections to Luther and Melanchthon, yet was careful in stating his position at this early time. Owing to

---

[1] Paul Schattenmann, <u>Die Einführung der Reformation in der ehemaligen Reichsstadt Rothenburg</u>, Einzelarbeiten zur Kirchengeschichte Bayerns, No. 7 (Munich: Kaiser, 1928), p. 63.

[2] He studied in Freiburg, Paris, at an Italian university, and in Basle, where he earned a doctorate.

[3] Hans-Joachim König, "Paul Speratus von Rötlen," <u>Ellwanger Jahrbuch</u> (1958-59), p. 81.

increasing pressure from high ecclesiastical circles, he left for Nuremberg in 1525.[1]

Ansbach

The preaching office at the St. Gumbertusstift in Ansbach was created in 1430 and was to be reserved for an educated person who either had a doctorate, licentiate, or bachelor's degree in the Holy Scripture. The preacher was supposed to deliver his sermons in German and teach at the <u>Stift</u>.[2] The patronage belonged to the margrave, as well as the deacon and the chapter. No one could act without the approval of the other.[3]

Georg Vogler, a politician, who was first private secretary to the chancellor of the principality, had extended personal conversations with Luther during the Diet at Worms in 1521. The effect of the meeting on Vogler and the margrave was so strong that the former even forged the margrave's decrees in order to advance Luther's ideas in Ansbach and the principality. His love for the "pure gospel" was accompanied by a strong hatred for the Pope and his "impious priests" (<u>Pfaffen</u>).[4]

Vogler had a close working relationship with Johann Rurer, since 1512 vicar at the <u>Gumbertusstift</u> and preacher at the same since 1528. In 1523 Rurer also became the town parson (<u>Stadtpfarrer</u>) at Ansbach's St. John. It is known that Vogler had sided with the new ideology in 1523 and consequently must have been an opinion leader.[5] But both Vogler and Rurer were the principal agents of religious change.

While the Lutherans had hoped that the margrave would lend stronger sup-

---

[1] Theodor Kolde, "Paul Speratus and Johann Poliander in Würzburg," <u>Beiträge zur Bayerischen Kirchengeschichte</u>, VI (1899), 65.

[2] Hermann Jordan, <u>Reformation und gelehrte Bildung in der Markgrafschaft Ansbach-Bayreuth</u>, Quellen und Forschungen zur Bayerischen Kirchengeschichte No. 1 (Leipzig: Heinsius, 1917), p. 32.

[3] Theobald Freudenberger, <u>Der Würzburger Domprediger Dr. Johann Reyss</u>, Katholisches Leben und Kämpfen im Zeitalter der Glaubensspaltung, No. 11 (1954), p. 25.

[4] Walter Brandmüller, "Dr. Johannes Winhart, der letzte katholische Stiftsprediger bei St. Gumbert in Ansbach," <u>Würzburger Diözesanblätter</u>, XVIII (1956), 129; and Elisabeth Grünewald, "Das Porträt des Kanzlers Georg Vogler (1550)," <u>Main-fränkisches Jahrbuch für Geschichte und Kunst</u> (1950), p. 135.

[5] Matthias Simon, "Johann Rurer," in <u>Religion in Geschichte und Gegenwart</u>, 3rd ed., V (Tübingen: Mohr, 1956), 1222.

port to their cause, the Catholics considered the phrase "to preach God's Word pure and plain" as a license and an excuse for Luther's religion. Rurer won the favor of the Ansbach population by abolishing the stole fees and mass stipends, thus decreasing the amount of revenue collected. While the magistracy and civil servants of the margrave supported Rurer, the margrave himself was cautious in his attitude toward the Protestants. The peasant uprising created more mistrust between the margrave and the cities. While Casimir's attention was concentrated on the unrest, Rurer passed out communion in both forms on Palm Sunday, 1525.[1] The condemnation of Rurer before the episcopal official and his subsequent rejection, led Casimir to lose the little sympathy he had for the new ideology.[2] The margrave ordered Vogler to be arrested, and Rurer fled to Silesia.

With the death of Margrave Casimir, the governors and magistracies (Räte) ordered Vogler released from jail, while Rurer returned from Silesia and was appointed to the preaching office in May, 1528.

Windsheim

Katherine Füchsly established a preachership in Windsheim in 1421. The occupant was to give sermons on Sundays and holidays and also on weekdays during the advent and feasting season. The position was especially desired because the preacher was freed from all contributions and taxes to the city and to the bishop. A later attempt by the bishop to collect a levy failed because of resistance from the magistracy. This office served as model for the one in Rothenburg. The occupant was required to have an academic degree. According to the deed of the foundation, the Teutonic Order paid 72 guilders, while the abbot of an adjacent monastery defrayed 15 guilders annually.[3]

In the first years the disciples of the new doctrine were associated mainly with the universities. However, a direct correlation between Windsheim students in Wittenberg and the new faith was not evident prior to 1522. The sudden enrollment of four men from this town at Luther's university in that year was a

---

[1] Karl Schornbaum, "Zur Reformation im Markgrafentum Brandenburg," Beiträge zur Bayerischen Kirchengeschichte, IX (1903), 26.

[2] Brandmüller, "Dr. Johannes Winhart," pp. 136-37.

[3] Johannes Bergdolt, Die freie Reichsstadt Windsheim im Zeitalter der Reformation 1520-1580, Quellen und Forschungen zur bayerischen Kirchengeschichte, No. 5 (Leipzig: Heinsius, 1921), pp. 11-12.

clear sign that the reformer's ideas had won adherents. The new ideas must have reached Windsheim from neighboring Nuremberg, where the small imperial city had strong political, economic and intellectual interests.[1] For example, it is known that Christoph Scheurl of Nuremberg sent Luther's writings to his many friends throughout southwestern Germany.[2]

When the preachership became vacant in January 1521, the magistracy requested Johann Herolt, who had been serving as the preacher at Nuremberg's New Hospital. But Herolt had to return to Nuremberg. In September, 1521, the magistracy asked a Leipzig university professor to recommend a qualified preacher from that city, or from Wittenberg. In November Georg Ebner, a student of Luther held his first sermon in Windsheim, but he, too, was unable to stay.

In the fall of 1522, the position was given to Thomas Appel, who had been expelled from the nearby Eichstätt bishopric for his Lutheran sermons. He enrolled in Wittenberg during the spring semester and became Prädikant there in October of 1522. Appel appealed mostly to the lower strata of society. His tumultuous, emotional appearances almost discredited the new movement among the more moderate followers of Luther. Appel, too, was forced out of office in early 1525, after having sided with the rebellious burghers of the town.[3] His appealing sermons led to the creation of a common treasury for the relief of the poor.[4] This gesture was welcomed by the local burghers.

Not surprisingly, the peasant uprising created a feeling of social distance between the city government and the lower strata, and Appel, the only potential appeaser, strongly sympathized with the latter. He was dismissed, but the protest raised by his supporters in favor of Appel was so strong that he was reinstated four weeks later. Even so, he was forced to leave town at the end of the peasant rebellion. Through the agency of Sleupner in Nuremberg, Andräus Altenstetter, an orthodox Lutheran became Appel's successor.[5]

---

[1] Ibid., p. 21.

[2] Hans J. Hillerbrand, "The Spread of the Protestant Reformation of the Sixteenth Century: A Historical Study in the Transfer of Ideas," South Atlantic Quarterly, Vol. XVII, No. 2 (Spring 1968), p. 274.

[3] Bergdolt, Die freie Reichsstadt Windsheim, pp. 27-28; and Matthias Simon, Pfarrerbuch der Reichsstädte Dinkelsbühl, Schweinfurth, Weissenburg in Bayern und Windsheim (Nürnberg, 1962), p. 89.

[4] Called "gemeiner Kasten" and first established in Leissnig, Saxony.

[5] Bergdolt, Die freie Reichsstadt Windsheim, p. 83.

Kitzingen

In Kitzingen, a small town along the busy road between Nuremberg and Frankfurt, a preachership was established in 1522. The mayor and the magistracy were its patrons. In October of 1521, Kitzingen's residents asked their margrave to grant them "evangelical freedom" and allow them to choose their own preacher. The choice fell on Christoph Hoffmann, a young man from Ansbach, who enrolled in Freiburg in 1513. He probably was associated with the Freiburg circle surrounding Engelbrecht and Zasius. In August of 1520, Hoffmann went to Wittenberg, where he earned a bachelor's degree in theology under Luther.[1] In August 1522, he started his Lutheran sermonizing in Kitzingen. He was soon joined by parson Martin Möglin. A number of traditional ceremonies were abolished, and the communion was offered in both forms in 1523.[2] When the provincial diet of Brandenburg-Ansbach passed a pro-Catholic resolution in October of 1526, both Hoffmann and Möglin, being married, were forced to flee. Hoffmann thereupon returned to Wittenberg.

Wertheim

Freudenberger mentions his uncertainty about the existence of a preaching foundation in Wertheim.[3] However, Simon ascribes to Jakob Strauss and definitely to Eberlin of Günzburg such a position in 1522 and 1526, respectively. Simon's assertion seems to be validated by the unusually high endowment placed upon this office. Count Georg promised not less than 100 guilders in gold when he sent a request for a clergyman to Luther in 1522.

Ulm and Upper Swabia

Ulm

Ulm was by far the largest city inside southwestern Germany, as well as the strongest imperial city in economic and political terms. Throughout medi-

---

[1] Matthias Simon, Ansbachisches Pfarrerbuch: Die Evangelisch-Lutherische Geistlichkeit des Fürstentums Brandenburg-Ansbach (1528-1806), Einzelarbeiten aus der Kirchengeschichte Bayerns, No. 27 (Nürnberg: Verein für bayerische Reformationsgeschichte, 1957), p. 204.

[2] Matthias Simon, Evangelische Kirchengeschichte Bayerns (München: Müller, 1942), I, 167, 169, 172.

[3] Freudenberger, Der Würzburger Domprediger Dr. Johann Reyss, p. 19.

eval times and well into the sixteenth century, it was the leading administrative headquarters among the Upper German imperial cities. This old trading center gradually evolved into an economically and politically powerful city-state not on the basis of its huge territory, but because of its excellent economic base. Its trade connections radiated out toward the east, west, south and north,[1] with its strongest ties in Upper Swabia and the interior part of Württemberg.[2]

The 55-year-old Franciscan, Eberlin von Günzburg, turned Lutheran in 1520 while living in Ulm. Soon afterwards he was transferred to Freiburg, from whence he engaged in missionary activity in a number of cities and towns of southwestern and central Germany as well as Switzerland. Eberlin's activity in Ulm was continued by his younger colleague Heinrich von Kettenbach, who began to preach Lutheran ideas in 1521.[3] Yet he and his sermons encountered such strong hostility from the readers and monks of the cloister that he soon left for Wittenberg.

Martin Idelhauser (Yedelhauser)[4] also began to defend Luther's teachings in 1521. As chaplain and foundation preacher at Ulm's parish church (the cathedral), he condemned all the traditional practices such as fasting, the mass, saint worship, penance, and purgatory as being false and superstitious. In July of 1522 he was near recanting before the bishop of Constance.

In 1523, Johann Diepold settled in Ulm after he had been forced to halt his evangelizing activity in Weil der Stadt. He carried on the reform movement in Ulm where Yedelhauser had left off. His popular sermons, free from bitter attacks on the Catholic Church and full of gentle appeal, produced a sharp increase in the number of adherents in the city. At the same time, the monk Jost Höflich, a native of Ulm, appeared as preacher. He had enrolled at Freiburg

---

[1] Karl S. Bader, Der deutsche Südwesten in seiner territorialgeschichtlichen Entwicklung (Stuttgart: Köhler, 1950), p. 156.

[2] This pattern is demonstrated by the distribution of the fairly wealthy Besserer clan. Their homebase was Ulm, but members could be found all over Upper Swabia (Memmingen, Ravensburg, Überlingen) and central Württemberg (Stuttgart, Cannstatt, Esslingen). See Heinrich Kramm, "Landschaftlicher Aufbau des deutschen Grosshandels," Vierteljahrschrift für Sozial- und Wirtschaftsgeschichte, XXIX (1936), 8, footn. 1.

[3] Paul Kalkoff, "Die Prädikanten Rot-Locher, Eberling und Kettenbach," Archiv für Reformationsgeschichte, XXV (1928), 132.

[4] A native Ulmer, who earned his B.A. in Vienna, his M.A. in Tübingen, and parson and permanent curate in nearby Urspring from 1520 to 1522. See also Albrecht Weyermann, Nachrichten von Gelehrten, Künstlern, und anderen merkwürdigen Personen aus Ulm (Ulm, 1798), p. 349.

University in 1515[1] and probably was a member of Philipp Engelbrecht's circle. After leaving the monastery, he passed through Ulm, sensing the desire of Ulm's inhabitants to learn about Luther's message. As he was not allowed to use one of the parish churches, he preached on the wooded heights surrounding the city. But the social unrest fostered by these gatherings encouraged the law-and-order minded magistracy in 1524 to capture and extradite Höflich to the bishop of Constance.[2] In 1525, he returned to Ulm, after rebelling peasants freed him from Meersburg jail.

The foregoing paragraphs have made one thing clear: the magistracy did not hesitate to voice its opposition to the ever-growing Reformation efforts within its walls. However, the growing movement soon resulted in a demand for freedom to proclaim the new faith through the means of an effective spokesman. The four representatives of the guilds asked their mayor for an evangelical preacher. He refused, but promised to bring the matter before the magistracy, which agreed that there was no choice but to comply. Immediately, the choice fell on Conrad Sam, a native of nearby Rottenacker and Lutheran preacher in Brackenheim until 1524. The man responsible for Sam's appointment was Sebastian Fischer, a nephew of Sam and chronicler in Ulm. After Sam had delivered three trial sermons, he was confirmed as the new preacher in Ulm's cathedral. His salary was set at 100 guilders. With Sam's installment, the number of converts in Ulm increased so rapidly that he was allowed to hire four more clergymen.

Sometime in 1525 or 1526, Sam seems to have favored Zwinglianism in the interpretation of the Lord's Supper. One reason for this change was his friendship with Johannes Oecolompadius, as well as his communications with other Zwinglian-oriented clergymen south of the Danube. However, the main reason seems to have been the strong pressure exerted by the mayor of Ulm, Bernhard Besserer, his son Georg, and their close friends Duke Ulrich and Prince Philipp of Hesse, all followers of Zwingli. In order to enjoy the necessary protection of the mayor against the vehement hostility of the Catholic monks and priests in the city, Sam shifted his allegiance to the Swiss reformer.[3]

---

[1] Cited by Mayer (Die Matrikel der Universität Freiburg, I, 219), as Jodocus Hofflich Ulmensis.

[2] Sigel, Das evangelische Württemberg, Vol. VII, No. 1, p. 86.

[3] Luther was informed about Sam's switch by Johann Schradin in Reutlingen and Andreas Althammer in Eltersdorf. See Albrecht Weyermann, "Die Bürger in Ulm, der zwinglischen Confession zugetan," Tübinger Zeitschrift für Theologie (1830), p. 142.

But Ulm's real shift toward the Protestant faith did not take place before 1528.[1] The patriciate was careful not to incur the wrath of the emperor and neighboring territories that could have had a negative impact on the city's trade and economic prosperity. Early in April of 1529, the city council expressed its determination to defend the gospel. In August of that year, it called on Philipp of Hesse to help draft a new church order (Kirchenordnung). Strasbourg advised Ulm to immediately institute a true evangelical liturgy, since religious freedom in the city had been suppressed for years.[2] The latter city responded in February of 1531 by confidentially informing Strasbourg about its plans to abolish mass, and asking the latter's advice on how to proceed.[3]

When in 1529 the second Diet of Speyer outlawed the Reformation movement, the evangelical faction protested, but its members were unable to agree to an armed alliance because of their different interpretation of the communion. Moreover, the intolerance of "the new pope" in Wittenberg disturbed Sam and Besserer, and the Ulm magistracy was forced to halt further changes, such as Sam's request to abolish the Latin mass. Nevertheless, Ulm now was to become deeply involved in the religious-political turmoil.

During the Augsburg Diet of 1530, Ulm took an equivocal stand, since it withheld support for the Lutheran as well as the Zwinglian doctrine.[4] It therefore saw itself as sitting between two chairs: the six cities (mostly Franconian) which signed the Lutheran Augsburg Confession, and the four Upper German (confessio tetrapolitana) cities of Strasbourg, Constance, Memmingen, and Lindau. Since 1529, these city-states between Constance and Ulm hoped for a partnership on the basis of the Zwinglian religion, which Ulm was expected to head. Yet Besserer, for personal reasons, seemed to shy away from any responsibility of defending such a pact in front of the Emperor. At this point, Strasbourg entered the picture, assuming leadership for a number of Upper German city-states.

---

[1] Heinrich Walther, "Bernhard Besserer und die Politik der Reichsstadt Ulm während der Reformation," Ulm und Oberschwaben, XXVII (1930), 11.

[2] Walther Köhler, Das Zürcher Ehegericht und Genfer Konsistorium, Quellen und Abhandlungen zur Schweizer Reformationsgeschichte, No. 10 (Leipzig: Heinsius Nachfolger, 1942), p. 12.

[3] Ibid., p. 42.

[4] "Die Bündnispolitik der oberdeutschen Städte des Schmalkaldischen Bundes unter dem Einfluss von Strassburg und Ulm 1529/31-1532" (unpublished Ph.D. dissertation, University of Tübingen, 1960), p. 35.

The demand by the Emperor to return to the Roman Catholic fold led the magistracy of Ulm to have the guildsmen of the city decide which side they preferred. As a result, 1576 citizens cast their votes for and 244 against the Reformation.[1] Sam's influence, which had by now reached its peak, opened the door for a further influx of Zwinglian ideas. In May of 1531, the magistracy appointed Bucer as well as Oecolompadius, Blarer, Bartholomäus Müller of Biberach, and Simprecht Schenk of Memmingen to carry out the reform movement in the entire city-state of Ulm. They first abolished mass, and then ordered the destruction of images and altars on June 19-20 of 1531. Zwingli's influence had its effect especially on church ceremonies, customs, and the church order, and less so in regard to the dogmatic dispute. Here, Ulm had to be considerate toward the Lutheran members of the Schmalkaldian Federation in the North. Nevertheless, the Ulm Reformation can be traced to the Zurich reformer.

Giengen

This imperial town was located in the Brenz Valley, also called the <u>Pfaffengasse</u> of Swabia.[2] This name came about owing to the numerous monasteries situated along the river and the complete encirclement by the territory of Heidenheim. Together with Riedlingen and Saulgau it had one of the oldest preaching foundations in Württemberg (founded 1420). All three institutions had the same donor, Johannes Mesner (also called von Riedlingen), the city doctor of Nuremberg. It must have been due to his frequent contacts and numerous acquaintances with local personalities on his trips between Nuremberg and Giengen that motivated Mesner to institute a preaching office there. Following the death of the founder, the magistracy and parson of Giengen assumed the right to nominate a preacher, which proved to be vital for the promulgation and diffusion of the Reformation.[3]

The <u>Prädikant</u> received a salary of 80 guilders plus a certain amount of natural produce.[4] Owing to the destruction of many documents from that time,

---

[1] The weavers guild voted 10 to 1, while the patricians voted only 2 to 1 against the Emperor. See <u>Beschreibung des Oberamts Ulm</u>, I, 89.

[2] <u>Beschreibung des Oberamts Ellwangen</u> (Stuttgart: Kohlhammer, 1886), p. 497.

[3] Rauscher, "Die ältesten Prädikaturen Württembergs," <u>Blätter für Württembergische Kirchengeschichte</u>, N.F., XXV (1921), 107-11.

[4] Andler, "Die Reformation in Giengen an der Brenz," <u>Blätter für Württembergische Kirchengeschichte</u>, N.F., I (1897), 100.

we know little about Giengen's development during the Reformation period. That the movement came comparatively late, in spite of Ulm's proximity may be partly blamed on its complete domination by the Heidenheim region.

The first evangelical sermon was given in 1528 by Caspar Pfeifelmann in the hospital church. Who he was, and where he studied is difficult to determine.[1] In the beginning of January, 1529, four Giengen burghers[2] petitioned their magistracy to appoint an evangelical preacher, an indication that the Reformation at last was coming off the ground.

Martin Rauber assumed the preaching post in the same year. He previously was chaplain in Esslingen, but when and how long he had held that position is not known. Neither do we know when Rauber left the Catholic Church, nor who influenced him to do so. In summer of 1531, he left Giengen to work for the movement in Ulm,[3] but returned again to that preachership in 1534.

Dillingen

In this temporary residence of the Augsburg bishops, the preaching office was commanded since 1518 or 1519 by Caspar Haslach. It was established through a donation from Bishop Heinrich von Lichtenau, who offered 1000 guilders, and was established by his follower, Bishop Christoph von Stadion. The right to nominate belonged to the episcopal city bailiff (Stadtvogt), and the presentation right to the magistracy. The prerequisite for the applicant was a doctorate or a licentiate in theology. In addition to the preaching office, Haslach received the benefice in Bernbeuren, which he staffed with a substitute.[4]

Luther's interrogation by Cajetan in Augsburg in October of 1518, and the conversion of the Augsburg preachers soon after, was bound to have an effect

---

[1] In 1534, Martin Frecht characterized Pfeifelmann (also called Miller) in the following words: negligence toward the Catholic Church and a mediocre education in the evangelical creed. Schiess, Der Briefwechsel der Brüder Thomas und Ambrosius Blaurer, I, 531.

[2] The four burghers were: Sixt Tuchhöffler, Mathias Kneuelin, Jakob Widenmann Kessler, and Peter Pfundstain. None of them could be found in the matriculation list of any university. Only a Christoph Pfundstain, probably Peter's brother, enrolled in Heidelberg in 1519. He very well could have been the initiator of the request. See Toepke, Die Matrikel der Universität Heidelberg von 1386 bis 1682, I, 518.

[3] Andler, "Die Reformation in Giengen an der Brenz," p. 100.

[4] Duncker, "Die Stellung des Prädikanten Kaspar Haslach zur Reformation," Zeitschrift für Bayerische Kirchengeschichte, XIV (1939), 132.

on the Danube communities. But the determining factor in Haslach's conversion was his close friend, Caspar Ammann, the preacher in neighboring Lauingen. Ammann, a prior of the Augustine monastery and an eminent Hebrew and Greek scholar, had displayed intense interest in Luther's writings since 1521. He influenced Haslach, his Hebrew student in that same year.[1] The two had long debates about the increasingly important ideology of the Wittenberg reformer.[2] Haslach also depended on Ammann for the supply of Luther's writings, which the latter got through Veit Bild in Augsburg.[3]

As it was, Haslach's first Protestant sermon in the Advent season of 1521 was based almost entirely on Luther's books Sermons for Advent (Adventspostille).[4] Early in 1522, Haslach was reprimanded by officials of the bishop, and in June of that year he was called to answer to charges of Lutheran heresy in Augsburg's ecclesiastical court. He, like Kress in Ellwangen, retracted from his earlier stand and in March of 1523 received the assignment as pastor in Bernbeuren.[5]

Munderkingen

In Munderkingen, the parish church was incorporated in the nearby cloister of Marchtal. The preacher was to be selected by both the town and the patron, Peter Mauler. The magistracy sent him to the Marchtal abbot, who, in turn, presented him to the bishop of Constance. The preaching office existed already in 1470. However, owing to its low endowment, the magistracy added a chapel to the office.[6] The exact extent of the endowment is not known.

In November, 1524, Paul Beck, a native of Munderkingen, took possession of the preachership. Whereas most Munderkingen students chose to study in Freiburg,[7] Beck, for some unknown reason, enrolled at Heidelberg Univer-

---

[1] Ibid., pp. 133-34.

[2] Alfred Schröder, "Der Humanist Veit Bild, Mönch bei St. Ulrich," Zeitschrift des Historischen Vereins für Schwaben und Neuburg, XX (1893), 180.

[3] Ibid., p. 210, No. 207.

[4] Duncker, "Die Stellung des Prädikanten Kaspar Haslach zur Reformation," p. 143.

[5] Ibid., pp. 139, 135, 157.

[6] Rauscher, "Die Prädikaturen in Württemberg," pp. 156, 159, 172, 179.

[7] Of 17 students between 1515 and 1524, 13 attended the University of Freiburg, 3 the University of Tübingen, and 1 the University of Heidelberg.

sity. He entered in 1516 and received his degree in 1523. Before his return to Munderkingen, he served as chaplain and helper in Heidelberg.[1] He probably had heard Luther's debate, and his positive impressions were reinforced through Brenz, Bucer, and other secondary reformers. He started holding Protestant sermons as soon as he returned to Munderkingen, and continued to do so until a full year after the Peasants' War in March, 1526. Beck enjoyed the popularity of his parish as well as the magistracy, who was particularly pleased with his calming role during the 1525 unrest. However, the Swabian Federation pressed the mayor and magistracy of Munderkingen to dismiss Beck. The remaining chaplain Rudolf Schnell continued the evangelical mission. After even this was brought to a halt, the citizens of Munderkingen went to Biberach and Ulm to attend sermons of the "seductive sect."[2]

Biberach

This city of approximately 3500 inhabitants was one of the eight larger towns in Upper Swabia. With Memmingen, Lindau, Ravensburg, and Überlingen, it was one of the most powerful cities economically. Most Biberachers attended the university in Tübingen between 1515 and 1525, followed by Freiburg; Heidelberg and Wittenberg trailed far behind.

With the establishment of the mass in Biberach's Holy Ghost Hospital, its priest was obliged to hold sermons for the hospital's poor and city residents.[3] Thus, the preaching office was originally a mere appendage to a mass benefice. Only at a later date did it attain a more independent status, although the tie with the hospital remained.[4] In 1491, the preaching office had already been transferred to the city's parish church.[5]

The first noticeable stirrings of the Lutheran movement in Biberach occurred in 1521, when a young but weird disciple of the Wittenberg reformer sought refuge in the house of Christoph Gräter, who was to become mayor in 1528. The name of this clandestine preacher was Schlupfindeck or "hide-in-the-corner." Following Gräter's instructions, he persuaded persons who were dis-

---

[1] Rauscher, "Die Prädikaturen in Württemberg," p. 204.

[2] Gustav Bossert, "Zur Geschichte des Evangeliums in Oberschwaben," Theologische Studien aus Württemberg, VII (1886), 35.

[3] Rauscher, "Die Prädikaturen in Württemberg," p. 55.

[4] Ibid., p. 173. [5] Ibid., p. 187.

satisfied with the Catholic Church to shed their allegiance. But the number of followers of the new faith grew slowly. Biberach's conservative population was mistrustful of the new movement, particularly since the early preachers were not from this town.[1] The few ardent Lutherans brought their friends to the meetings so that the secret gatherings eventually reached a stage where they evolved into public meetings. Schlupfindeck was succeeded by Strohschneider, who probably came from Memmingen. During the week, Protestant sermons were held by Allgäuer and Bayer, as well as by Red Hans, names which imply their place of origin.[2] The magistracy assumed a neutral wait-and-see attitude, but soon after 1525 took the Protestant point of view. A Memmingen bookseller sold Lutheran literature on market days. Following the conversion of Schappeler, Memmingen became the focal point for the entire Reformation movement in Upper Swabia, especially Biberach.

After Christoph Gräter became mayor in 1528, the magistracy seized the jurisdiction over the local clergy from the bishop. He had the strong support of some guild masters. Now the city government was able to staff the ecclesiastical positions with men of their choice. Since a number of zealous Reformation sympathizers considered Luther's teachings to be lacking in socio-political vigor and decisiveness, they turned toward Zurich and Strasbourg for support. They admired the Wittenberg reformer's religious stamina, but considered him too conservative on social and political issues.

After 1525, the influence of Ulm grew stronger and gradually replaced that of Memmingen. The preacher, Bartholomäus Müller from Ulm, was given a preaching office in 1527. Three years later, Martin Cless of Göppingen was granted the office by the magistracy. He was first recommended to Zwingli and Oecolompadius by Duke Ulrich's secretary, who, in turn, probably proposed Cless for Biberach, via Ulm.

As soon as Oecolompadius, Bucer, and Ambrosius Blarer had introduced the Swiss church order in Ulm and its surroundings, they visited Biberach on their way home. With the abolition of the mass, and the destruction of the images in April and June of 1531, the Protestant movement went ahead full steam. From these facts it can be gathered that Biberach's reform movement was strongly influenced first from Memmingen, and later from Ulm.

---

[1] Albert Angele, ed., <u>Altbiberach um die Jahre der Reformation</u> (Biberach: Renner, 1962), p. 183.

[2] Hans Maier from Kempten, Allgäu, enrolled at the University of Vienna in 1516, and 1518. Willy Szaivert and Franz Gall, eds., <u>Die Matrikel der Universität Wien</u> (Wien: Universitätsverlag, 1959), II, 436.

Memmingen

In 1479, the merchant family by the name of Vöhlin established a preachership and perpetual mass at St. Martin's Church in Memmingen. The occupant was required to present a doctorate or licentiate, or a bachelor's degree in Holy Scripture or canon law. The presentation right stayed with the oldest member of the Vöhlin family, in this case Erhard Vöhlin of Illertissen. If he should die, the mayor and magistracy would assume the task of presenting the candidate to the bishop of Augsburg.[1]

In 1520, another preaching foundation was installed at the parish of St. Elizabeth. The mayor and the magistracy of Memmingen demanded that the occupant be either a lay priest or a member of the convent.[2]

The preachers at St. Martin and the Elizabethan nunnery, Christoph Schappeler and Christoph Gerung, both preached the new doctrine in 1523. Schappeler, a native of the Swiss town of St. Gallen, and a former monk, had a solid education.[3] He also had a good knowledge of the Holy Scripture. His doctorate in theology and licentiate in the law fully met the tough requirements of this office. Prior to his appointment at St. Martin, Schappeler had served ten years as school master in St. Gallen. With his appointment in 1514, he immediately began agitating, charging the well-to-do with exploiting the poor. His concern for the disadvantaged, which he expressed through inflammatory sermons, soon made Schappeler popular. But he aroused the ire of the conservative members of the magistracy, among them Vogelmann, the town's syndicus.[4] He had close personal ties to the school master in St. Gallen, Gebentinger, and the foundation preacher of that town, Dr. Joachim Vadian. Gebentinger had served as go-between to Schappeler and Zwingli in Zurich. In 1522, Schappeler left the city to go to Switzerland, either to be near Zwingli or to seek a preachership. But he soon returned. Vadian forwarded his 67 final speeches (Schlussreden) to Schappeler in the beginning of 1523.[5] And in November of that year,

---

[1] Wolfgang Schlenck, Die Reichsstadt Memmingen und die Reformation, Memminger Geschichtsblätter (1968), pp. 23-24.

[2] Ibid.; and Friedrich Dobel, Memmingen im Reformationszeitalter (Memmingen: Besenfleder'schen, 1877), p. 35.

[3] Schappeler received his education at the University of Leipzig, where he also taught between 1505 and 1510.

[4] Askan Westermann, Eberhart Zangmeister (Memmingen: Renner, 1932), p. 21.

[5] Schappeler himself sat besides Vadian and Hofmeister on the second debate in Zurich.

he delivered the first "Lutheran" public sermon in St. Martin.[1] The violent counterattack by Catholic priests, including the threat to indict Schappeler before the bishop, almost led to his return to Switzerland.[2] Zwingli already had tried to secure a recently vacated preaching office in Winterthur for Schappeler, but the Memmingen magistracy encouraged him to stay.[3]

Soon a circle of like-minded persons, such as the patricians Hans Ehinger, Dr. Matthäus Neidhard, Laux Pfister, Raphael Sättelin, and members of the various guilds came into being. Among the latter, Sebastian Lotzer, then a merchant and formerly a furrier, was particularly committed to the new cause. During the Peasants' War he had commanded the Baltringen group in the rural area between Memmingen and Ulm. The Latin and German school masters Paul Hopp and Johann Schmidlin, and Memmingen's second preacher, Gerung, lent Schappeler their support.[4]

Because of the weak backing the Catholics received from the magistracy, the reform movement was able to progress smoothly. Beginning in December of 1524, communion in both forms was served in St. Martin's. The clergy were at least put on equal footing with the rest of the Memmingen burghers. Mass was abandoned, and communion was made mandatory.[5]

In January, 1525, Memmingen hired an additional preacher, Simprecht Schenk. While on his way home from his parish outside Zurich, he stopped in Memmingen and was asked by Schappeler to deliver several sermons. Immediately he was offered the preachership at Our Lady's Church.[6] This victory of the Reformation had its impact on the imperial city of Kaufbeuren, whose magistracy held a public debate in January of 1525. It was similar to the debate held in Memmingen a week earlier.

During the occupation by troops of the Swabian Federation in summer of 1525, the Memmingen movement experienced a severe setback. Schappeler

---

[1] Schlenck, Die Reichsstadt Memmingen, p. 31.

[2] Dobel, Das Reformationswerk zu Memmingen unter dem Drucke des Schwäbischen Bundes 1525-29 (Augsburg: Lampart, 1877), p. 29.

[3] Ibid., p. 30.

[4] Westermann, Eberhart Zangmeister, p. 31.

[5] Schlenck, Die Reichsstadt Memmingen, p. 43.

[6] Emil Schenck, Simprecht Schenck: Das Lebensbild eines schwäbischen Reformators (Darmstadt: Selbstverlag, 1938), p. 4.

fled to St. Gallen, while Schenk was fired. The city was now without an evangelical preachership. By popular demand however, the magistracy hired the Swiss Lutheran Georg Gugy as preacher a few months later. Together with Michael Hilsbach, he was one of the very few preachers south of the Lutheran-Zwinglian border who refused to renounce the idea of a real presence in the communion. Both had close contact with Capito in Strasbourg. Meanwhile, the demand for a second preacher in Memmingen forced the magistracy to request the services of Dr. Johann Wanner of Constance. The latter was recommended to Memmingen by Georg von Frundsberg of nearby Mindelheim. His objective was to strengthen and solidify the Reformation movement in central Upper Swabia, by coupling the movements in two neighboring towns.[1]

After a short stay in Memmingen, Wanner continued his evangelizing activities in Mindelheim, but returned to Memmingen in January, 1527. Meanwhile, Schenck was reinstated in 1526 at St. Martin. In the ensuing quarrel between Gugy and Schenck, Wanner consistently defended the Zwinglian-oriented Schenck. In order to settle the dispute between Lutherans and Zwinglians once and for all, and to further the Reformation, Ambrosius Blarer arrived in Memmingen in November, 1528. Memmingen was so pleased with his performance that it asked Blarer to remain in town, but the Constance magistracy required his return after a three-month stay.

So far, we have observed a shift in geographical orientation of Memmingen's citizens from St. Gallen to Constance, which took place with the escape of Schappeler and prior to the occupation of the city by the forces of the Swabian Federation in 1525. From now on, the Constance magistracy and its two preachers Wanner and Blarer took over the reform efforts in Memmingen and Mindelheim, the first time that Constance could function as diffusion center for other German cities.

Isny

This small imperial town enjoyed a lively trade with the Upper Swabian cities of Constance, Augsburg, and Nuremberg. Its close affinity with Nuremberg, Memmingen, Kempten, and St. Gallen may also have aided in the spread of Protestant ideas in the town.[2] It is in this light that we must see the abbot's

---

[1] Dobel, Das Reformationswerk zu Memmingen, p. 29.

[2] Immanuel Kammerer, "Die Reformation in Isny," Blätter für Württembergische Kirchengeschichte, LIII (1953), 4, 5.

complaint about the merchant Peter Buffler and the town clerk, Hans Voelk, as "the true foundation of the Lutheran faction."

In 1462 or earlier, the Constance canon Hans Guldin, together with the magistracy and citizens of Isny, donated 600 guilders for a preaching office, demanding that the occupant hold sermons on Sundays and holidays.[1] The patronage lay with the mayor and magistracy. The salary was approximately 65 guilders annually,[2] 30 guilders coming from Guldin, the rest was donated by the town.[3]

The first Protestant occupant of the preachership was Conrad Frick. He assumed the position in 1518, and changed to the evangelical creed in 1525. Where Frick received his education and motivation to adopt this new position is unknown. The merchant Peter Buffler had close connections to Nuremberg where his brother notified him of the changes in the Reformation movement.

It is owing to Frick, for example, that the parson Steudlin commenced to serve communion under both forms on Easter, 1525. The resistance from the Catholics and a possible military retaliation against the Reformation movement in Isny demanded the unification of the Protestant cities of Upper Swabia. In January, 1529, the Isny mint master complained to Zwingli about indifference from Zurich, expressing the hope that Constance might lend its support.[4] Like the rest of Upper Swabia, Isny had been under the influence of the Swiss Reformation, especially since the Bern debate in 1528, in which the town was represented by Paul Fagius. A fellow student of Brenz, Schnepf, and Billican in Heidelberg, he became rector of the local Latin School in 1527. Later, after he finished his studies and teaching in Strasbourg, he succeeded Frick in 1537.

On his return from Zurich in 1528, Fagius delivered Frick an encouraging letter from Zwingli. Fagius himself exchanged letters with the Zurich reformer at a later period. To what extent Zurich served as a model for Isny's Reformation movement is hard to tell, but the reading of the Bible in Hebrew, Greek, and Latin that was begun in the Isny parish church, reminds one of a similar practice in Zurich. When the Lutheran and Zwinglian preachers of Kempten held a debate in Isny to settle their differences in 1529, Frick, Fagius,

---

[1] Rauscher, Die Prädikaturen in Württemberg, p. 167.

[2] Ibid., p. 190.

[3] Kammerer, "Die Reformation in Isny," p. 78.

[4] Ibid., p. 13.

and the assistant Wolfgang Gasser sided with Jakob Haystung, the Zwinglian preacher.[1]

Kaufbeuren

In Kaufbeuren, the early Reformation movement can be traced to the influence of Memmingen and Constance. Luther became subject for discussion in this imperial town when he visited Augsburg in 1518, and so was the public announcement of the Edict of Worms in 1521.[2]

The occupant of the preaching office, Jakob Lutzenberger, had expounded Lutheran ideas since 1521. This led to a quarrel with the parson at St. Martin and the Hospital Church. The preaching office at the latter church was established in 1453 by Ulrich Honold and the magistracy and was better endowed than all the regular benefices.[3] An additional preaching office seems to have been installed in the early 1520's at St. Martin, and occupied by Dr. Sebastian Fuchssteiner. He also read Luther's writings to the parish.[4]

Of much greater influence on the citizens was Lutzenberger. How he came to preach the new ideas is not exactly known. That he had close links to Johann Wanner seems probable. Wanner had met Luther in Erfurt in the years 1506-1509, and he preached in Kaufbeuren between spring 1520 and 1521. Whether or not he confessed to the new faith at this early time is unknown. In May of 1524, Kaufbeuren acquired a Protestant mayor, Blasius Honold, who, among other things, permitted German mass and communion in both forms. However, owing to the town's precarious geographical location between the Catholic states of Waldburg and Bavaria, Honold personally asked neighboring towns such as Memmingen and Kempten about their measures against the old church. When the magistracy called for a debate between the town's Catholic and Protestant factions, it demanded the protocol of a similar debate held in Memmingen one week before.

---

[1] Ibid., pp. 15-16.

[2] M. Weigel, "Der erste Reformationsversuch in der Reichsstadt Kaufbeuren und seine Niederwerfung," Beiträge zur Bayerischen Kirchengeschichte, XXI (1914), 149.

[3] Karl Alt, Reformation und Gegenreformation in der freien Reichsstadt Kaufbeuren, Einzelarbeiten aus der Kirchengeschichte Bayerns, No. 15 (Munich: Kaiser, 1932), p. 10, footn. 6.

[4] Weigel, Die Deutschordenskomturei in Rothenburg, p. 153.

Christoph Schappeler was asked to preside and deliver an introductory guest sermon.[1] Schappeler was forced to cancel his appearance, however, and Wanner came from Constance. Meanwhile, adjacent Kempten had fully endorsed Kaufbeuren's plans for a religious debate but warned of possible riots.[2] In the debate, Lutzenberger sharply condemned the use of idols and images. The magistracy drew the conclusion that mass and other man-made ceremonies were contrary to the Bible, but planned to abolish these only after seeking further advice from the other cities. Wanner was immediately commissioned to contact Ambrosius Blarer, while word was expected from Augsburg. Blarer and Wanner, in a joint statement, opposed the removal and destruction of images in favor of a more gentle measure: persuasion by sermons. They advised Kaufbeuren to continue the traditional mass, provided that it not be considered an offering, and that the low mass be eliminated. Under pressure from the Swabian Federation and neighboring Bavaria in the Peasants' War, Kaufbeuren had to cancel not only any intended changes, but with the occupation of the town by forces of the Swabian Federation in May-June, 1525, the evangelical preachers Fuchssteiner and Lutzenberger were expelled.

Kempten

The citizens and magistracy of Kempten desired to be independent from the rich and powerful local Benedictine monastery. Since its abbot was not interested in the creation of a preaching position, the city did so by itself in 1474. The monastery also opposed a Latin School, but the city finally won the approval of the Pope and the emperor. Now most of the pupils were sent to this school instead of the monastery school.[3]

The foundation preacher at St. Mang, Caspar Heelin[4] was also the opinion leader of Kempten's Protestant community. He acquired this post in 1508. Not only did he influence Sixt Rummel, then parson at St. Mang,[5] but also the dea-

---

[1] Ibid., pp. 193-94. [2] Ibid., p. 184.

[3] Otto Erhard, Die Reformation der Kirche in Kempten (Kempten: Danneheimer, 1917), pp. 1-4.

[4] Also mentioned as M. Paul Hölin. He was from Herrenberg, enrolled at the universities in Tübingen and Freiburg between 1482 and 1487/88, and was parson in Rottenburg from 1499 to 1505.

[5] Otto Erhard, "Die Sakramentsstreitigkeiten in Kempten 1530-1533," Beiträge zur Bayerischen Kirchengeschichte, XVII (1911), 153.

con, M. Jakob Haystung, who likewise joined the movement in 1520. Haystung was a graduate of Freiburg university in 1518, and later became chaplain of the Holy Ghost Hospital, and in 1523 was preacher-helper at St. Mang. In 1527, Haystung became the preacher of the foundation for one year. The curate at the collegiate church of St. Lawrence, Matthäus Waibel, a peasant's son and graduate of the University of Vienna, openly preached the new doctrine in 1523 or earlier. He and Haystung were friends,[1] reinforcing each other in their opposition to the practices within the Roman Catholic hierarchy.

Waibel supported Rummel in his attempts to bring about religious change, and in February, 1525, Rummel proudly labelled Waibel "our Allgäu heretic."[2]

Following the defeat of the peasants, the Swabian Federation tried to capture Waibel, but he found refuge in Caspar Heelin's house while continuing his Protestant sermons at St. Mang. Yet on August 26, 1525, a group of armed men from the Swabian Federation lured him out of Kempten and hanged him in neighboring Leutkirch.[3]

In February, 1525, Rummel and his chaplains shed their priestly garbs, performed the baptismal ceremony in German, eliminated candles, the mass for the dead, and discontinued the custom of fasting.

The "Big Buy" (grosse Kauf) on May 6, 1525, terminated the subservient status of the city to the abbot and gave the signal for radical changes in the liturgy. Monstances, chalices, and relic containers[4] were melted into coins to pay the abbot. Finally, the mass[5] was abolished.[6]

Beginning in 1527/28, Kempten began to feel the influx of Zwinglian ideas from Switzerland, creating a rift between Protestants. Wolfgang Maler, a schoolmaster from the local Latin School, was a follower of the Zwinglians. But it was especially Jakob Haystung who stood under the Swiss influence. As long as Sixt Rummel was parson, however, conflict was prevented, although he favored the Lutheran communion. With Rummel's death in 1529, the magistracy asked the cities of Augsburg and Nuremberg for a "learned and able per-

---

[1] Otto Erhard, Matthias Waibel: Treu dem Evangelium, No. 5 (1525), p. 4.

[2] Erhard, "Die Sakramentsstreitigkeiten in Kempten," p. 154.

[3] Erhard, Matthias Waibel, pp. 10-11.

[4] Reliquienbehältnisse.

[5] Called the "hochwürdige Sacrament."

[6] Erhard, "Die Sakramentsstreitigkeiten," p. 154.

son" to succeed Rummel. Since Augsburg was found leaning toward the Zwinglians and Nuremberg toward the Lutherans, these inquiries give us a clue to the opposite tendencies within the Kempten church hierarchy and the city government. Nuremberg suggested that Kempten notify Luther personally concerning the need of a parson.

As the confessions of the Zwinglians and Tetrapolitana towns[1] had no chance of being accepted on the Augsburg Diet in 1530, Kempten officially remained in contact with Nuremberg until 1532, and was therefore labeled "Lutheran." Owing to Haystung, however, a shift toward the Zwinglian side took place during these two years.[2] The magistracy tried to compromise between the opposing Reformed and Lutheran parties by proposing that communion be administered individually, according to choice. But the two Lutheran preachers, Johann Serranus and Johann Rottach rejected this proposition and pointed to Kempten's endorsement of the Confessio Augustana two years earlier, a fact recognized by the city fathers.[3]

To find a solution to this problem, Kempten asked Bucer in Strasbourg to mediate between the groups. Both agreed to Bucer's formula, but when the Lutheran side again reversed its stand, the magistracy expelled Serranus and Rottach and replaced them with Zwinglian preachers from Switzerland.[4] The Zwinglian school, represented by Haystung, had already scored a victory when 500 against 174 Kempten citizens voted for the removal of images in January, 1533, shortly before the expulsion of the two preachers.[5]

Summary

(1) Preaching foundations, or preacherships, most of which were set up during the late Middle Ages, played an important role in the Reformation.

(2) The emphasis of Luther on the sermon, and the excellent education of their occupants made these foundations susceptible to the new ideology. Southwestern Germany was an area characterized by a profusion of such foundations.

---

[1] Constance, Lindau, Memmingen, and Strasbourg.

[2] Erhard, "Die Sakramentsstreitigkeiten," p. 160.

[3] Ibid., p. 164.

[4] Ibid., p. 169; Haystung's letter to A. Blarer on February 1, 1533. See Schiess, Briefwechsel der Brüder Ambrosius und Thomas Blaurer, I, 382.

[5] Erhard, "Die Sakramentsstreitigkeiten," p. 170.

Many of the influential reformers held these preacherships, hence cities such as Schwäbisch-Hall, Strasbourg, Ulm, and Constance were important diffusion centers marked by a great stability.

(3) However, the foundations must be seen within the context of the total Reformation movement. The latter did not usually spread from one preachership to another, but was dependent on receptive communities, irrespective of any existing preaching foundations. Many of the foundations adopted the new ideology somewhat later than ordinary heads of parishes.

(4) Since the demands made on preachers in respect to proficiency were unusually high, applicants could not be limited to a narrow geographic range, as was often the case with common parsons. Instead, they were frequently hired from distant cities. Sometimes magistracies were in direct consultation with universities, which supplied many of the foundation preachers.

(5) Finally, the preaching foundations determined the nature and character of the church service in southwestern Germany for centuries to come.

CHAPTER VIII

CONCLUSIONS

The diffusion of a new religious ideology, namely that of the evangelical movement within southwestern Germany between 1518 and 1534, has been the subject of this investigation. This movement marked the beginning of one of the most deeply felt tremors experienced in the Western World, one that ultimately led to fundamental and important transformations in the ecclesiastical, social, political, and cultural spheres. This period was characterized by changes that emanated strictly from within the clergy, with the consent of their parishes. City and territorial leadership, the knights excepted, did not play an assisting role prior to the years 1524-1525. It is the purpose of this last chapter to restate the major findings of the study and relate them to the findings of diffusion studies undertaken by Hägerstrand, Hudson, and others.

From its spiritual-religious inception, the Reformation movement in Germany went through four successive developmental stages: (1) It began with the diffusion of the interesting news that Luther was an opponent of the Roman-Catholic establishment. However, the real Reformation began with (2) the preaching of Luther's ideas. This was then followed by changes in the liturgy, such as (3) the communion under both forms, and, finally, by (4) the abolition of mass, which symbolized a total break with the Roman-Catholic Church and the beginning of the political Reformation. These four steps represent almost a full circle, beginning with disassociation from the existing ecclesiastical institution and ending as a full-fledged institution in itself.

Maps showing the place and date of the first evangelical sermon, as well as the two liturgical changes were prepared in order to see the spatial spread and importance of events that are judged to be key criteria of the Reformation movement. These maps demonstrate the spontaneous and widespread initial acceptance of Luther's ideas, as well as the subsequent interventive actions by a number of localities that were able to prevent and reverse the changes in the existing liturgy. Needless to say, had these political authorities not intervened, all four stages of the Reformation would probably have been accepted in time over the entire German Southwest.

Of the four successive developmental stages, stage 2, the place and date of the first evangelical sermon, was most decisive, since it represented the most characteristic and typical criterion of the movement. It was and remained the nexus of the entire Reformation, pointing to a number of other changes. Since the new sermon was based on the Word, it negated the hierarchical position and power of the Pope and the bishops over that of the local parsons and preachers. It symbolically did away with the traditional division into an all-dominant clergy and subservient laiety and also ended the extension of the hierarchical-ecclesiastical jurisdiction over the secular realm.

A look at the regional-cultural characteristics of the German Southwest gives some indication of the decentralized nature of the political and cultural map that encouraged outside influences emanating from Switzerland and the Upper Rhine area, centering on Zürich, Basel, Strasbourg, and Constance, on one hand, and central Germany on the other hand. The lack of important cultural and economic nuclei in the central part of our area, Württemberg, largely explains the division of this duchy and the entire Southwest into Lutheran and Zwinglian-oriented territories, a division that runs north and south of Stuttgart. That this religious divide survived the political Reformation in Württemberg in 1534 is an indication of the comparative cultural and political weakness of this territory.

A number of important roads and thoroughfares channeled the Reformation into specific sections of the area. Although direct relationships between these communication routes and the influence and movement of preachers are not easy to trace, there are, nevertheless a few that stand out. It should come as no surprise that most of the evidence for such links appear in the southern section of Upper Germany, whose urban realms experienced the Reformation more intensely and directly than the northern section. In the latter region the change emanated largely from such educational centers as Heidelberg and Wittenberg, as well as from persuasive territorial rulers.

Natural barriers such as the Black Forest, seem to have influenced the spread of evangelical ideas in certain directions. The resulting irregular spatial pattern of distribution in this highland area helped to channel the ideas of the Strasbourg reformers in a north-south direction. Simultaneously, the important Kinzig Valley thoroughfare functioned as a ready valve, stimulating the towns therein as well as Villingen and Rottenburg on the east side of the range. The Swabian Forest (Schwäbischer Wald) adjoining the Rems Valley to the north likewise had a definite barrier effect. Sernatingen (today's Ludwigshafen) and

Ravensburg could also have been a result of the barrier effect of Lake Constance. At the same time Constance, on the southwest side of the lake, was not only large enough, but also had the proper personal resources to make itself felt on a number of cities north and east of the lake. Where politically-imposed boundaries prevailed as a constraining influence, no similar effect can be detected. Of course, this barrier effect can be noticed only in situations where a powerful commercial and ecclesiastical central place was located near a physical-cultural barrier. It is safe to assume that strong and far-reaching commercial and social connections were an important prerequisite for overcoming any type of barrier.

Archival sources of the <u>Hauptstaatsarchiv</u> in Stuttgart permit the comparison of the number of taxpayers to non-taxpayers in 27 prefectures in Württemberg. The analysis contests the Marxist assertion that the early Reformation was an innovation carried mainly by the economically disadvantaged section of the population. On the contrary, all available evidence points to the opposite conclusion: the more prosperous prefectures were more disposed to accept the new ideology and demand an evangelical preacher.

The ecclesiastical Reformation of the sixteenth century was the result of a revised theological concept. Not unlike today, theology was practiced exclusively within universities. Hence, by far the most important agents for the diffusion of evangelical ideas into the public domain were the universities, where political constraints on Luther's teachings were non-existent at least during the early phase of the movement. The universities in central Germany, as well as the universities in Heidelberg and Freiburg acted as important diffusion centers at a very early date. Their impact on the religious map of the 1520's can be determined largely by their spheres of influence.

Following political harassment by authorities of the new ideology, opinion leaders from Heidelberg and Freiburg relocated themselves in other cities relatively close by. Strasbourg and Schwäbisch-Hall, therefore, emerged as new diffusion centers, with Strasbourg influencing such smaller centers as Stuttgart, Esslingen, Rottenburg, and Horb. Schwäbisch-Hall, by virtue of its devoted reformer Johannes Brenz, became <u>the</u> diffusion center for the northern region of southwestern Germany. The Zwinglian/Bucerian conflict afflicting this area was successfully resolved in favor of the Wittenberg reformer thanks to this city. Besides procuring preachers for the neighboring cities and towns, Schwäbisch-Hall assumed the responsibility of acting as adviser to them. Memmingen, Ulm, Biberach, Kempten, and Lindau, on the other hand, relied on St.

Gallen, Constance, Augsburg, and Basle as sources of advice on reform and how and when changes should take place. As the Reformation progressed toward the final break with the Roman-Catholic Church, the reliance on these centers increased, leading to a certain standardization of formerly diverse views and practices.

The diffusion of a newly conceived ideology, be it religious or political, is more complicated than the diffusion of a material item such as automobiles, telephones, and even customary habits. Although both are dependent on propagators located in diffusion centers of varying importance, they differ in several respects. Unlike the individual ownership of material goods, an ideology can be the common property of a group of people. And unlike its material counterpart, the spread of an emerging religious ideology is far more difficult to observe and keep track of. Its ultimate success depends to a large degree on political authority acting as an intervening variable that transforms such movements into a dynamic process in which localities and territories are conquered or lost. In the case of the Reformation, the split within the Protestant ranks into Lutherans, Zwinglians, and a number of other groups was an additional variable that must be taken into account. This split weakened the two major Protestant parties in their contest with the Roman Catholic Church for the religious-spatial domination of the German Southwest. Finally, it must be noted that the diffusion of religious values, unlike that of material ones, is less bound to geographical distance and space.

The spread of the evangelical ideology was in fact most closely bound to personal relationships, the principal means of familiarizing the broad majority of the German people--90 percent of whom were then illiterate--with Luther's teachings. As has been shown, these personal relationships either could be long standing and pre-date the origin of the innovation, or they could be more recent personal ties that originated with the innovation. It stands to reason that both were equally important.

A comparison of the findings in this work with those of modern diffusion studies presents several difficulties. For example, Hägerstrand and other scholars have had at their disposal quantified information, such as the number of persons possessing an automobile or telephone. Data of a similar nature, such as the existing number and location of proselytes of the new faith around the second decade of the sixteenth century are, of course, unavailable. Therefore, it was necessary in this study to rely on the date and locality of the appearance of the first preachers and/or opinion leaders who proclaimed the new religion. Any persuasion and diffusion process that depends on single persons has an erratic and sporadic quality that is difficult to fit into generalized diffusion schemes. Although the vast majority of non-resident and resident preachers assumed their positions at the request of the population whom they served, it is unknown how strong the movement was at any particular time.

Notwithstanding the many unknown variables, one can match up the spatial movement and distribution of preachers with any one of the following types of diffusion: (1) expansion or contagious diffusion, (2) relocation diffusion, and (3) hierarchical diffusion.[1] All three types operated simultaneously and in varying intensities during the first fifteen years of the Reformation. Though one must keep in mind that viewing diffusion purely from a spatial angle gives an incomplete picture. It tends to neglect the personal and institutional connections that were and are so important in the spread of an ideology. No diffusion study should ignore these.

Expansion diffusion, or the gradual spread of an idea from a center outward, occurred primarily in the initial phase of the movement and applied mostly to the spreading of rumors among the population about Luther's opposition to the existing Roman-Catholic Church hierarchy. It was spontaneous, unmanipulated, and hardly encountered any resistance among the general population. The term "modified expansion diffusion," e.g., a mixture of expansion and hierarchical diffusion, can be applied to the function of the preachers in the northeastern corner of our territory. Like the vast majority of preachers in northern Germany, they were graduates and friends of Luther at Wittenberg. The "spillover effect" of the Zurich-based movement on several upper German towns and villages along the Rhine between Basle and Constance can be traced to the same type of diffusion. The impact of cities like Esslingen, Ulm, and Kempten on smaller, mostly rural settlements surrounding them, must also be included here.

Relocation diffusion, involving a change of the preacher's domicile took place primarily because of constraints imposed by territorial, ecclesiastical, and city authorities on the new religion. It implied neither an increase in the number of preachers, nor additional reformed communities. The negative effect of such movements was felt primarily in the Duchy of Württemberg, where most cities and towns lost their preachers <u>in toto</u> either to independent imperial city-states within or outside their realm, or to territories in Franconia and

---

[1] John C. Hudson, <u>Geographical Diffusion Theory</u>, Northwestern University Studies in Geography, No. 19 (Evanston: Department of Geography, 1972).

along the upper Rhine. For example, most of the preachers hired by the city-state of Ulm in 1527 and later had been relocated from neighboring towns as well as more distant localities in Württemberg, Baden, and Switzerland. It has been shown that in some cases these authoritative constraints put only a temporary constraint on the movement, and that communities deprived of their preachers were receptive to it as soon as the governmental limitations were removed. Where no such censorship was operative and the migrating preacher was a member of a larger body of newly-oriented clergy that remained in their positions, one can speak of relocation diffusion only in a limited sense. It is worthwhile to note that the relocation was often achieved through an agent. This was always a city or town of considerable size, which handled requests coming from smaller towns or villages within its sphere of influence. It is characteristic of this type of movement that expansion and relocation diffusion not only met but also complemented each other. In such cases, the city that acted as the relocation agent and the locality from where the preacher was expelled, were in different areas.

Hierarchical diffusion operated all over our territory. In fact, it appears to have been not only the dominant, but also the most important type of diffusion. This should come as no surprise, since during its first fifteen years of Reformation movement was limited mostly to cities and towns. Southwestern Germany was the most urbanized section and possessed the Empire's largest cities. In general, the larger the city, the readier it was to accept the ideology and spread it into the countryside. The primary center of Wittenberg and Sickingen's Ebernburg are notable exceptions.

The correlation between city size and zone of influence can be observed best by comparing the range of Strasbourg and Augsburg with that of most medium-size cities. The radius of influence of the former was much larger, although Heidelberg, Worms, Freiburg, and Constance commanded a wide range as well, nearly matching that of Strasbourg and Augsburg, and surpassing that of Ulm, the fourth-largest metropole in the Southwest. The reason behind this anomaly must be sought with the propagators of the Reformation who were active in these places. In Heidelberg, Worms, and Augsburg, Luther himself appeared in front of the clergy, academicians, students, and political leaders. Constance had a number of educated and devoted opinion leaders, plus well-established political and commercial ties with cities in Upper Swabia. In contrast Ulm, as dominant metropole and trading center of southwestern Germany, had no qualified evangelical leaders.

Another form of hierarchical (but non-spatial) diffusion was operative, that is, a sociological diffusion: Luther's ideas were conveyed from and by persons in higher status levels to those in lower ones, thereby slowly but steadily increasing the number of adherents. Hierarchical diffusion can also imply that certain cities functioned as opinion leaders for other, mostly smaller cities and communities. It was in these larger metropoles that Luther's arguments were first heard and thoroughly examined.

The problem with the concepts of expansion, relocation, and hierarchical diffusion is their limited applicability to a number of the cases in this study. For example, Luther came to the Worms Diet in 1521, where he converted a number of political leaders. Upon their return to their respective territories, they requested an evangelical preacher either from Wittenberg or a larger city close by. The type of diffusion applicable to this process is difficult to determine.

In regard to the personal qualifications and characteristics of the preachers or innovators, the great majority were of younger age than Luther himself. This fact is contrary to the opinion of political scientists who attach great interest and importance to political processes if only generated by persons of higher relative age.[1]

The study of the adoption rate among the very early sympathizers, i.e., those preaching prior to 1520, shows that on the average they were younger than their successors. This corroborates findings in other studies.[2] For example, during the years 1521 to 1525, one can detect a balance tending toward the older age brackets, although the majority of preachers still were of Luther's age or younger. This seems to confirm the theory that the Reformation was largely a conflict between generations. It appears to have been a rebellion on the part of the younger generation against the older. One of the persistent characteristics of a revolution is that its advocates are young, and hence in conflict with an older generation that is skeptical toward them and their ideas.[3]

If we trace the proportion of evangelical missionaries along a vertical axis, and the year of adoption along a horizontal axis (Figs. 17 and 18), we can

---

[1] Paul F. Lazarsfeld, Bernard Berelson, and Hazel Gaudet, The People's Choice (New York: Columbia University Press, 1955), p. 44.

[2] Everett M. Rogers, Diffusion of Innovations (New York: The Free Press, 1962), p. 172.

[3] Moeller, "Die deutschen Humanisten und die Anfänge der Reformation," Zeitschrift für Kirchengeschichte, LXX (1959), 56-57.

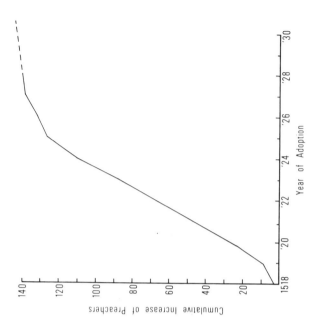

Fig. 18.--Cumulative Adoption Curve of Luther's Ideas 1518-1530.

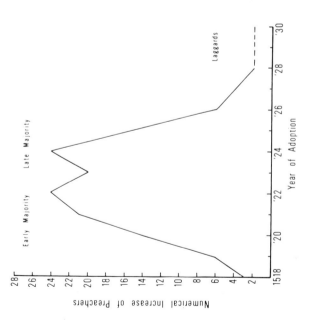

Fig. 17.--Adoption Rate of Preachers 1518-1530.

detect some fascinating parallels and discrepancies in the adoption pattern of more mundane items. Both conform to the logistic of an S-shaped curve. However, while the latter usually experiences a slow growth at the beginning due to resistance, the Reformation movement was marked by a relatively swift progress in its first two years, 1518-1519. This rapid rise can be explained in part by the fact that Luther did not plan a complete break with the Catholic Church, but rather tried to improve the system from within, an objective shared by many people. In spite of their rapid acceptance, Luther's ideas were to become even more popular in the ensuing three years. The year 1523 suddenly experienced a decline in the growth process, which, however, was made up in the following year. The two years between 1524 and 1526 registered a drop as sharp as the rise experienced five years earlier, with again a fairly swift decline after that. The drop can be interpreted (1) as a result of the numerous rebellions that took place in several cities in 1523, a harbinger of the more violent insurgency two years later called the Peasants' War, and (2) as a consequence of the vastly declining number of students who attended the universities during that period.

It might be asked why did the evangelical movement continue in the face of the declining number of believers? It might be argued in return that the movement was sustained by the same persons who had adopted the new faith during the preceding eight years. Nevertheless, this decline of active supporters began to make itself felt in the early 1530's and was manifested in the shortage of available preachers.[1] This can be considered the prime reason why the majority of new preachers in cities and towns were non-indigenous after 1525-1526. Finally, while many social and religious movements enter an era of consolidation and simplification, the Reformation split up instead into several groups that displayed great animosity toward each other.

This study further shows that a two-step flow hypothesis is applicable to the diffusion of the evangelical faith. Influences stemming from either the mass media or direct personal contact with Luther first converted a large group of pioneers or opinion leaders, including clergymen, university professors, teachers, and doctors. Under the term "mass media" should be understood all means of communication, both written or printed, as well as the information conveyed by word-of-mouth. It made the majority of people aware of

---

[1] Letter written by A. Blarer to the mayor and magistrat of Memmingen on November, 1531. See Traugott Schiess, ed., *Der Briefwechsel der Brüder Thomas und Ambrosius Blaurer*, 3 vols. (Freiburg: Fehsenfeld, 1908-12), I, 284-85.

the new movement. Still, genuine adoption of the new faith depended on the persuasion of recognized leaders and friends who were able to offer information and advice. The combination of these two processes resulted in a multiple-step flow in which certain individuals became converts and then active propagators of the innovation.

The maps of this study give only a limited understanding of the diffusion process and are inadequate because they leave out the "why and how" underlying the propagation of the Reformation movement. In order to answer these questions, it was necessary to study the inter-personal relationships between major and local reformers. These relationships, as well as a consideration of the social, economic, political, institutional, and cultural matrix in which the Reformation took place, were found to be essential for understanding of the spatial development of the Reformation movement in southwestern Germany.

BIBLIOGRAPHY

Published Material

Adam, Johann. Evangelische Kirchengeschichte der Elsässischen Territorien bis zur Französischen Revolution. Strassburg: Heitz, 1928.

Aich, Johann. Geschichte des Marktortes Laupheim. Blaubeuren: H. Baur, 1913.

Aland, Kurt. "Die Theologische Fakultät Wittenberg und ihre Stellung im Gesamtzusammenhang der Leucorea während des 16. Jahrhunderts." Kirchengeschichtliche Entwürfe. Gütersloh: Mohn, 1960. I, 283-394.

Albert, Peter P. "Die reformatorische Bewegung zu Freiburg bis zum Jahre 1525." Freiburger Diözesan-Archiv, N. F. XLVI (1919), 1-80.

Allgemeine Deutsche Biographie. Leipzig: Duncker & Humblot, 1875.

Alt, Karl. Reformation und Gegenreformation in der freien Reichsstadt Kaufbeuren. Einzelarbeiten aus der Kirchengeschichte Bayerns, Vol. XV. München: Kaiser, 1932.

Amman, Hektor. "Oberdeutsche Kaufleute und die Anfänge der Reformation in Genf." Zeitschrift für Württembergische Landeskunde, XIII (1954), 150-93.

Andler, --. "Die Reformation in Giengen an der Brenz." Blätter für Württembergische Kirchengeschichte, N. F. I (1897), 97-113.

Andreas, Willy. "600 Jahre Reichsstadt Gengenbach." Zeitschrift für die Geschichte des Oberrheins, CVIII (1960), 292-306.

Angele, Albert, ed. Altbiberach um die Jahre der Reformation. Biberach: Renner, 1962.

_____. "Zur Geschichte des Bergbaues von Durlach von 1439 bis 1532." Zeitschrift für die Geschichte des Oberrheins, I (1850), 43-48.

Auer, Paul. Geschichte der Stadt Günzburg. Günzburg: Donauverlag, 1963.

Bachteler, Kurt. Geschichte der Stadt Gross-Sachsenheim. Bietigheim: Glaser, 1962.

Bader, Josef. "Aus der Geschichte des Pfarrdorfes Griessen im Klettgau." Freiburger Diözesan Archiv, IV (1869), 232-44.

Bader, Karl S. Der deutsche Südwesten in seiner territorialgeschichtlichen Entwicklung. Stuttgart: Koehler, 1950.

———. "Die oberdeutsche Reichsstadt im Alten Reich." Esslinger Studien, XI (1965), 25-38.

———. "Reichsadel und Reichsstädte in Schwaben am Ende des alten Reiches." Festschrift zum 70. Geburtstag von Theodor Mayer. Konstanz: Thorbecke, 1962. I, 248-63.

Barton, Peter F. "Das Jahr 1525 und die Abschaffung der Messe in Strassburg." Reformation und Humanismus, Robert Stupperich zum 65. Geburtstag. Martin Greschat [ed.]. 1969. Pp. 141-57.

Batzer, Ernst. "Neues über die Reformation in der Landvogtei Ortenau sowie in den Städten Gengenbach und Offenburg." Zeitschrift für die Geschichte des Oberrheins, N. F. XXXIX (1926), 63-74.

Baum, Adolf. Magistrat und Reformation in Strassburg. Strassburg: Heitz, 1887.

Baumann, Ludwig. "Zur schwäbischen Reformationsgeschichte. Urkunden und Regesten." Freiburger Diözesan-Archiv, X (1896), 97-124.

Beck, Paul. "Die Reformation in Riedlingen und ihr Herold." Württembergisches Vierteljahresheft für Landesgeschichte, IV (1895), 170-75.

Bender, Karl. "Die Reformation in Gengenbach." Beiträge zur Badischen Kirchengeschichte. Veröffentlichungen des Vereins für Kirchengeschichte in der evangelischen Landeskirche Baden, Vol. XXII. 1962. Pp. 1-19.

Bergdolt, Johannes. Die freie Reichsstadt Windsheim im Zeitalter der Reformation (1520-1580). Quellen u. Forschungen zur Bayerischen Kirchengeschichte, Vol. V. Leipzig: Deichert, 1921.

Bertogy, Hercli. Beiträge zur mittelalterlichen Geschichte der Kirchengemeinde am Vorder- und Hinterrhein Chur. Chur: Bischofberger, 1937.

Böhringer, Wilhelm. Heimatbuch Reichenbach an der Fils. Reichenbach: Schaul & Schniepp, 1968.

Bossert, Gustav. "Aktenmässige Leidensgeschichte einer evangelischen Gemeinde Württembergs." Blätter für Württembergische Kirchengeschichte, I (1886), 4-12.

———. Aus Horb a. N. und Umgebung. Horb a. N.: Christian, 1936.

———. "Beiträge zur badisch-pfälzischen Reformationsgeschichte." Zeitschrift für die Geschichte des Oberrheins, LVI (1902), 37-89, 250-90, 401-49.

———. "Beiträge zur Geschichte der Reformation in Franken." Theologische Studien aus Württemberg, I (1880), 173-212.

———. "Briefe und Akten zur Geschichte der fränkischen Reformatoren." Theologische Studien aus Württemberg, VII (1886), 1-28.

Bossert, Gustav. "Die Jurisdiktion des Bischofs von Konstanz 1520-1529." Württembergische Vierteljahreshefte für Landesgeschichte, N.F. XI (1893), 260-81.

———. "Die Syngrammatisten." Blätter für Württembergische Kirchengeschichte, VII (1892), 19-21.

———. "Drei Haller Biographien." Württembergisch Franken, VII (1900), 75-76.

———. "Johann Herolt." Württembergische Vierteljahreshefte für Landesgeschichte, IV (1881), 289-96.

———. "Johann Isenmann." Blätter für Württembergische Kirchengeschichte, VII (1901), 137-49.

———. "Kleine Beiträge zur Geschichte der Reformation in Württemberg." Blätter für Württembergische Kirchengeschichte, N.F. VIII (1904), 144-80.

———. "Vorwärts." Blätter für Württembergische Kirchengeschichte, V (1890), 95.

———. "Zur Brenzbiographie." Blätter für Württembergische Kirchengeschichte, X (1906), 97-116.

———. "Zur Geschichte der Reformation in Esslingen 1522 und 1523." Blätter für Württembergische Kirchengeschichte, N.F. VIII (1893), 92-94.

———. "Zur Geschichte des Evangeliums in Oberschwaben." Theologische Studien aus Württemberg, VII (1886), 28-51.

———. "Zur Geschichte Stuttgarts in der ersten Hälfte des sechzehnten Jahrhunderts." Württembergisches Jahrbuch (1914), pp. 138-81; 183-243.

Boulding, Kenneth. "The Learning Process in the Dynamics of Total Societies." The Study of Total Societies. Ed. Samuel Z. Klausner. Garden City: Anchor, 1967. Pp. 98-113.

Brandmüller, Walter. "Dr. Johannes Winhart, der letzte katholische Stiftsprediger bei St. Gumbert in Ansbach." Würzburger Diözesanblätter, XVIII (1956), 112-36.

Brunner, F. "Historische Notizen über die Pfarrei Ballrechten." Freiburger Diözesan-Archiv, XIV (1881), 283-97.

Buchholz, E. W. Ideologie und latenter sozialer Konflikt. Göttinger Abhandlungen zur Soziologie, Vol. XV. Stuttgart: Enke, 1968.

Buck, Hermann. Die Anfänge der Konstanzer Reformationsprozesse, Österreich, Eidgenossenschaft und Schmalkaldischer Bund 1510/22-1531. Tübingen: Osiandersche, 1964.

Burkhardt, Georg. Geschichte der Stadt Geislingen an der Steige. Konstanz: Thorbecke, 1963.

Burkhardt, R. "Paulus Beck der erste evangelische Geistliche Geislingens." Blätter für Württembergische Kirchengeschichte, XXXV (1931), 249-65.

Burmeister, Karl H. Achilles Pirmin Gasser. 1505-1577. Wiesbaden: Steiner, 1970.

―――――. Thomas Gassner. Lindau: Museumverein, 1971.

Calwer Verlagsverein, ed. Württembergische Kirchengeschichte. Calw and Stuttgart: Vereinsbuchhandlung, 1893.

Cantz, Max. "Caspar Kantz und die Nördlinger Reformation." Historischer Verein für Nördlingen und Umgebung, XII (1929), 143-74.

Clauss, Herman. Die Einführung der Reformation in Schwabach 1521-1530. Quellen und Forschungen zur Bayerischen Kirchengeschichte, Vol. II. Leipzig: Deichert'sche, 1917.

Cless, David F. Versuch einer kirchlichen politischen Landes- und Culturgeschichte von Württemberg bis zur Reformation. Schwäbisch Gmünd: Lanz, 1808.

Conrad, Otto. "Johann Gayling von Ilsfeld, ein Reformator Württembergs." Blätter für Württembergische Kirchengeschichte, N.F. XLIII (1939), 13-27.

Crisman, Miriam Usher. Strasbourg and the Reform. New Haven and London: Yale University Press, 1967.

Deutsch, Karl W. The Analysis of International Relations. Foundations of Modern Political Science Series. Englewood Cliffs: Prentice Hall, 1968.

Dobel, Friedrich. Das Reformationswerk zu Memmingen unter dem Drucke des Schwäbischen Bundes 1525-1529. Augsburg: Lampart, 1877.

―――――. Memmingen im Reformationszeitalter nach handschriftlichen und gleichzeitigen Quellen. Memmingen: Besenfelder'schen, 1877.

Duncker, Ludwig. "Die Stellung des Prädikanten Kaspar Haslach zur Reformation." Zeitschrift zur Bayerischen Kirchengeschichte, XIV (1939), 129-59.

Egger, Joseph. Geschichte Tirols. Innsbruck: Wagner, 1880.

Egli, Emil. Schweizerische Reformationsgeschichte. Zürich: Zürcher & Furrer, 1910.

Enders, Ernst, ed. Dr. Martin Luther's Briefwechsel. Frankfurt: Evangelischer Verein, 1884.

Erhard, Otto. Der Bauernkrieg in der gefürsteten Grafschaft Kempten. Kempten: Kosel, 1908.

―――――. "Die Sakramentsstreitigkeiten in Kempten 1530-1533." Beiträge zur Bayerischen Kirchengeschichte, XVII (1911), 153-73.

―――――. Matthias Waibel. Kempten and München: Kosel'schen, 1925.

Ernst, Fritz. "Zur Geschichte Schwabens im späteren Mittelalter." Festschrift für Karl Bohnenberger 75. Geburtstag. Ed. Hans Bihl. Tübingen: Mohr, 1938. Pp. 76-81.

"Evangelisches Leben in Geislingen vor der Reformation." Geschichtliche Mitteilungen von Geislingen und seiner Umgebung, IV (1933), 1-11.

Fehn, Klaus. Die zentralörtlichen Funktionen früher Zentren in Altbayern. Wiesbaden: Steiner, 1970.

Foerstemann, Carolus E., ed. Album Academiae Vitebergensis 1502-1560. Leipzig: Tauchnitz, 1841.

Franz, Günther. Der Deutsche Bauernkrieg. 4th ed. Darmstadt: Gentner, 1958.

Freudenberger, Theobald. Der Würzburger Domprediger Dr. Johann Reyss. Katholisches Leben und Kämpfen im Zeitalter der Glaubensspaltung, Vol. XI. Münster: Aschendorff, 1954.

Gagliardi, Ernst; Müller, Hans; and Büsser, Fritz, eds. Johannes Stumpfs Schweizer- und Reformationschronik. Basel: Birkhäuser, 1952.

Gall, Franz, ed. Die Matrikel der Universität Wien. Wien: Universitätsverlag, 1959.

Gaisburg-Schöckingen, Friedrich von. "Das Rottenburger Wappenbuch." Reutlinger Geschichtsblätter, XVIII (1907), 1-24.

Gebhardt, Bruno, ed. Handbuch der deutschen Geschichte, 8th ed. Stuttgart: Union, 1955.

Giefel, Andreas. "Zur Geschichte des Andreas Wendelstein von Rottenburg." Reutlinger Geschichtsblätter, XIV (1903), 32-35.

Glitsch, Cristoph. "Die Bündnispolitik der oberdeutschen Städte des Schmalkaldischen Bundes unter dem Einfluss von Strassburg und Ulm, 1529/31-1532. Dissertation, Tübingen, 1960.

Gmelin, Julius. Hällische Geschichte. Schwäbisch Hall: Staib, 1896.

Gothein, Eberhard. Wirtschaftsgeschichte des Schwarzwaldes und der angrenzenden Landschaften. Strassburg: Trübner, 1892.

Götz, Johann B. Die Primizianten des Bistums Eichstätt aus den Jahren 1493-1577. Reformationsgeschichtliche Studien und Texte, Vol. DXIII. Münster: Aschendorff, 1934.

Gradmann, Robert. Süddeutschland. Stuttgart: Engelhorns, 1931.

Grube, Walter. "Aus der Geschichte von Stadt und Amt Güglingen." Zeitschrift des Zaubergäuvereins (1958), pp. 49-57.

_____. "Stadt und Amt in Altwürttemberg." Stuttgarter Zeitung. June 29, 1972, p. 35.

Grube, Walter. "Württembergische Verfassungskämpfe im Zeitalter Herzog Ulrich." Neue Beiträge zur Südwestdeutschen Landesgeschichte, XVII (1962), 144-60.

Hagenmaier, Winfried. Das Verhältnis der Universität Freiburg i. Br. zur Reformation. Dissertation, Freiburg, 1968.

Hägerstrand, Torsten. Innovation Diffusion as a Spatial Process. Translated by Allen Pred. Chicago: University of Chicago Press, 1967.

Haisch, Andreas. Der Landkreis Mindelheim in Vergangenheit und Gegenwart. Mindelheim: Kreistag und Landratsamt, 1968.

Haller, Johannes. Die Anfänge der Universität Tübingen. Stuttgart: Kohlhammer, 1929.

Hartmann, Julius, and Jäger, Karl. Johannes Brenz. Hamburg: Perthes, 1842.

Haselier, Günter. "Die Auswirkungen der Territorialisierung auf die politisch-kulturelle Struktur in der Pfalz und im unteren Elsass." Probleme der Geschichte und der Landeskunde am linken Oberrhein. Ed. Franz Artehn. Freiburg: Veröffentlichungen des Instituts für Oberrheinische Landesgeschichte, 1966. Pp. 84-88.

_____. Geschichte des Dorfes und der Gemeinde Weiher am Bruhrain. Weiher: Selbstverlag, 1962.

Heckel, Martin. "Reformation: Rechtsgeschichtlich." Evangelisches Staatslexikon. Ed. Herman Kunst, et al. Stuttgart: Kreuz, 1966. Pp. 1804-33.

Hermelink, Heinrich, ed. Die Matrikel der Universität Tübingen. Stuttgart: Kohlhammer, 1931.

_____. Die theologische Fakultät in Tübingen vor der Reformation. Tübingen: Mohr, 1906.

_____. "Luthertum der württembergischen und der bayerischen Landeskirche." Festgabe für Landesbischof D. Theophil Wurm zum 80. Geburtstag. Stuttgart: Quell, 1948. Pp. 149-54.

Herold, Reinhold. Geschichte der Reformation in der Grafschaft Oettingen 1522-1569. Schriften des Vereins für Reformationsgeschichte, Vol. LXXV. Halle: Niemeyer, 1902.

Herrmann, Franz. Markgrafen-Büchlein. Bayreuth: Mühl, 1902.

Herrmann, Rudolf. "Die Prediger im ausgehenden Mittelalter und ihre Bedeutung für die Einführung der Reformation im Ernestinischen Thüringen." Beiträge zur Thüringischen Kirchengeschichte, I (1929-1931), 20-68.

Hillerbrand, Hans J. "The Spread of the Protestant Reformation of the 16th Century." South Atlantic Quarterly, LXVII (1968), 265-86.

Hitchcock, William R. The Background of the Knights' Revolt 1522-1523. University of California Publication in History, Vol. LXI. Berkeley and Los Angeles, 1958.

Hoffmann, Gustav. "Reformation im Bezirk Besigheim." Blätter für Württembergische Kirchengeschichte, XXXVIII (1934), 133-204.

_____. "Reformation und Gegenreformation im Bezirk Welzheim." Blätter für Württembergische Kirchengeschichte, XIV (1910), 29-36.

Hoffmann, Joseph E. "Michael Stifel." Esslinger Studien, XIV (1968), 21-37.

Holborn, Hajo. "The Social Basis of the German Reformation." The Reformation. Material or Spiritual? Ed. Lewis W. Spitz. Problems in European Civilization. Lexington: Heath, 1962. Pp. 37-42.

Hölzle, Erwin. Beiwort zu: Der deutsche Südwesten am Ende des alten Reiches. Stuttgart: Kohlhammer, 1938.

Hopf, A. "Hans Jakob Wehe. Erster lutherischer Pfarrer in Leipheim." Beiträge zur Bayerischen Kirchengeschichte, II (1896), 145-59.

Hudson, John C. Geographical Diffusion Theory. Northwestern University Studies in Geography, Vol. XIX. Evanston: Department of Geography, 1972.

Iserloh, Erwin. Der Kampf um die Messe. Katholisches Leben und Kämpfen im Zeitalter der Glaubensspaltung, Vol. X. Münster: Aschendorff, 1952.

Jäger, Karl. Mitteilungen zur schwäbischen und fränkischen Reformationsgeschichte. Stuttgart: Kohlhammer, 1828.

Jordan, Hermann. Reformation und gelehrte Bildung in der Markgrafschaft Ansbach-Bayreuth. Quellen und Forschungen zur bayerischen Kirchengeschichte, Vol. I. Leipzig: Deichert'sche, 1917.

Jung, F. Johann Schwebel. Kaiserslautern: Kayser, 1910.

Jung, Wolfgang. "Zur Geschichte des evangelischen Gottesdienstes in der Pfalz." Veröffentlichung des Vereins für Pfälzische Kirchengeschichte, VII (1959), 1-11.

Kähni, Otto. "Reformation und Gegenreformation in der Reichsstadt Offenburg und Landvogtei Ortenau." Die Ortenau, XXX (1950), 20-37.

Kalkoff, Paul. "Die Prädikanten Rot-Locher, Eberlin und Kettenbach." Archiv für Reformationsgeschichte, XXV (1928), 128-50.

_____. "Die Stellung der deutschen Humanisten." Zeitschrift für Kirchengeschichte, XLVI (1928), 161-231.

Kammerer, Immanuel. "Die Reformation in Isny." Blätter für Württembergische Kirchengeschichte, LIII (1953), 3-64.

Kantzenbach, Friedrich. "Theologie und Gemeinde by Johannes Brenz." Blätter für Württembergische Kirchengeschichte, LXV (1965), 3-38.

Kaser, Kurt. "Zur politischen und sozialen Bewegung im deutschen Bürgertum des 15. und 16. Jahrhunderts." Deutsche Geschichtsblätter, III (1902), 1-18; 49-60.

Kattermann, Gerhard. Die Kirchenpolitik Markgraf Philipp I. von Baden (1515-1533). Veröffentlichung des Vereins für Kirchengeschichte in der evangelischen Landeskirche Badens, Vol. XI. Lahr: Schauenburg, 1936.

Kaufmann, Alexander. "Einige Bemerkungen über die Zustände des Landvolks in der Grafschaft Wertheim während des 16. und 17. Jahrhunderts." Freiburger Diözesan-Archiv, II (1866), 47-60.

Kaufmann, Georg. Geschichte der deutschen Universitäten. Stuttgart: Cotta, 1896.

Kawerau, G. "Johannes Draconites aus Carlstadt." Beiträge zur Bayerischen Kirchengeschichte, III (1897), 247-75.

Keidel, Friedrich. "Johannes Piskatorius." Blätter für Württembergische Kirchengeschichte, VI (1902), 143-78.

_____. "Ulmische Reformationsakten von 1531 und 1532." Württembergische Vierteljahreshefte für Landesgeschichte, N. F. XIV (1895), 255-342.

Keim, Theodor. Schwäbische Reformationsgeschichte bis zum Augsburger Reichstag. Tübingen: Fries, 1855.

Keller, Siegmund. "Der Adelsstand des süddeutschen Patriziates." Festschrift Otto Gierke zum 70. Geburtstag. Weimar: Böhlau, 1911. Pp. 741-58.

Kerker, M. "Die Predigt in der letzten Zeit des Mittelalters mit besonderer Beziehung auf das südwestliche Deutschland." Theologische Quartalschrift, XLIII (1861), 373-410.

Keyser, Erich, ed. Badisches Städtebuch. Stuttgart: Kohlhammer, 1959.

_____. Bayerisches Städtebuch. Teil I: Franken. Stuttgart: Kohlhammer, 1967.

_____. Württembergisches Städtebuch. Stuttgart: Kohlhammer, 1962.

Kläui, Paul. "Rottweil und die Eidgenossenschaft." Zeitschrift für Württembergische Landesgeschichte, XVIII (1959), 1-14.

Klaus, Bernhard. "Herkunft und Bildung lutherischer Pfarrer der reformatorischen Frühzeit." Zeitschrift für Kirchengeschichte, LXXX (1969), 22-49.

Kobe, Fritz. Die erste lutherische Kirchenordnung in der Grafschaft Wertheim. Veröffentlichungen des Vereins für Kirchengeschichte in der evangelischen Landeskirche Badens, Vol. VII. Lahr: Schauenburg, 1933.

Köhler, Walther. Zürcher Ehegericht und Genfer Konsistorium. Quellen und Abhandlungen zur schweizerischen Reformationsgeschichte, Vol. X. Leipzig: Heinsius, 1942.

_____. Zwingli und Luther, Vol. I. Leipzig: Heinsius, 1924.

Kohls, Ernst-Wilhelm. "Evangelische Bewegung und Kirchenordnung in oberdeutschen Reichsstädten." Zeitschrift der Savigny-Stiftung für Rechtsgeschichte, Kan. Abt., LXXXIV (1967), 110-34.

Kohls, Ernst-Wilhelm. Evangelische Bewegung und Kirchenordnung. Veröffentlichungen des Vereins für Kirchengeschichte in der evangelischen Landeskirche Badens, Vol. XXV. Lahr: Schauenburg, 1966.

Kolde, Theodor. Andreas Althamer der Humanist und Reformator in Brandenburg-Ansbach. Erlangen: Junge, 1895.

──────. D. Johannes Teuschlein und der erste Reformationsversuch in Rothenburg o. d. T. Erlangen: Böhme, 1901.

──────. "Die Berufung des Kaspar Greter als Stiftsprediger nach Ansbach." Beiträge zur Bayerischen Kirchengeschichte, V (1899), 197-226.

──────. "Zur Geschichte Billicans und Althammers und der Nördlinger Kirchenordnung vom Jahre 1525." Beiträge zur Bayerischen Kirchengeschichte, X (1904), 28-40.

König, Hans-Joachim. "Die Freundschaft zwischen Johannes Brenz und dem Crailsheimer Pfarrer Adam Weiss." Württembergisch Franken, N. F. LV (1971), 82-93.

──────. "Paul Speratus von Rötlin." Ellwanger Jahrbuch (1958/59), pp. 81-85.

Krebs, Manfred. "Politische und kirchliche Geschichte der Ortenau." Die Ortenau, XVI (1929), 116-58.

──────. Die Protokolle des Speyerer Domkapitels. Veröffentlichung der Kommission für geschichtliche Landeskunde in Baden-Württemberg, Reihe A, Vol. XXI. Stuttgart: Kohlhammer, 1952.

Kuhn, Werner. Die Studenten der Universität Tübingen zwischen 1477 und 1534. Göppingen: Helmbrecht, 1971.

Lauppe, Ludwig. "Die Reformation im klösterlich-schwarzwäldischen Kirchspiel Scherzheim-Lichtenau." Die Ortenau, XXXII (1952), 71-84.

Lazarsfeld, Paul F.; Berelson, Bernard; and Gaudet, Hazel. The People's Choice. 2nd ed. New York: Columbia University Press, 1948.

Lederle, Karl F. "Die kirchlichen Bewegungen in der Markgrafschaft Baden-Baden." Freiburger Diözesan-Archiv, XLV (1917), 367-450.

Lehmann, Hartmut. "Luther und der Bauernkrieg." Geschichte in Wissenschaft und Unterricht, XX (1969), 129-39.

Leube, Ed. Im Zeichen von Sankt Christoph. Urach: Brenz, 1928.

Liermann, Hans. "Laizismus und Klerikalismus des evangelischen Kirchenrechts." Zeitschrift der Savigny-Stiftung für Rechtsgeschichte, Kan. Abt. LXX (1953), 1-27.

Luther, J. "Vorbereitung und Verbreitung von Martin Luther's Thesen." Greifswalder Studien zur Lutherforschung und neuzeitlichen Geistesgeschichte, Vol. VII. Berlin: Gruyter, 1933.

Marte, J. D. Die Auswärtige Politik der Reichsstadt Lindau von 1530 bis 1532. Beilage zum Jahresbericht der Kgl. Realschule Ludwigshafen/Rh. XIX. 1904/05.

Mauersberg, Hans. Wirtschafts- und Sozialgeschichte Zentraleuropäischer Städte in neuerer Zeit. Göttingen: Vandenhoeck & Ruprecht, 1960.

Mayer, Hermann, ed. Die Matrikel der Universität Freiburg im Breisgau von 1460 bis 1656. Freiburg: Herdersche, 1907.

Meyer von Knonau, G. "Die eidgenössische Besatzung in der Reichsstadt Lindau im spanischen Erbfolgekrieg." Schriften des Vereins für die Geschichte des Bodensees, XLIV (1915), 39-47.

_____. "Zürcherische Beziehungen zur Reichsstadt Lindau." Schriften des Vereins für die Geschichte des Bodensees, XLI (1912), 3-13.

Merkel, Friedemann. Geschichte des Evangelischen Bekenntnisses in Baden von der Reformation bis zur Union. Veröffentlichung der Verfassung für Kirchengeschichte in der evangelischen Landeskirche Badens, Vol. XX. Lahr: Schauenburg, 1960.

Michel, Lothar. Der Gang der Reformation in Franken. Erlanger Abhandlungen zur mittleren und neueren Geschichte, Vol. IV. Erlangen: Palm & Enke, 1930.

Moeller, Bernd. "Die deutschen Humanisten und die Anfänge der Reformation." Zeitschrift für Kirchengeschichte, LXX (1959), 46-61.

_____. "Die Kirche in den evangelischen freien Städten Oberdeutschlands im Zeitalter der Reformation." Zeitschrift für die Geschichte des Oberrheins, CXII (1964), 147-62.

_____. Johannes Zwick und die Reformation in Konstanz. Gütersloh: Mohn, 1961.

_____. Reichsstadt und Reformation. Schriften des Vereins für Reformationsgeschichte, Vol. CLXXX. Gütersloh: Mohn, 1962.

Mone, F. I. "Predigerpfründe im 14. und 15. Jahrhundert zu Heidelberg, Lahr und Basel." Zeitschrift für die Geschichte des Oberrheins, XVIII (1865), 1-11.

Müller, Gottfried. "Reformation und Information." Die Bedeutung der Reformation für die Welt von Morgen. Ed. Rainer Schmidt. Frankfurt: Lembeck, 1967. Pp. 198-207.

Müller, Walter. Die Stellung der Kurpfalz zur lutherischen Bewegung von 1517 bis 1525. Heidelberger Abhandlungen zur mittleren und neueren Geschichte, Vol. LX. Heidelberg: Winter, 1937.

Münch, Ernst. Geschichte des Hauses und Landes Fürstenberg. Aachen: Mayer, 1829.

Näf, Werner. Vadian und seine Stadt St. Gallen. St. Gallen: Fehr'sche, 1957.

Neu, Heinrich. Geschichte der evangelischen Kirche in der Grafschaft Wertheim. Heidelberg: Winter, 1903.

_____. Pfarrerbuch der evangelischen Kirche Badens von der Reformation bis zur Gegenwart. Lahr: Schauenburg, 1938.

Oberamtsbeschreibungen:

    Beschreibung des Oberamts Brackenheim. Stuttgart: Lindemann, 1873.

    Beschreibung des Oberamts Crailsheim. Stuttgart: Kohlhammer, 1884.

    Beschreibung des Oberamts Ellwangen. Stuttgart: Kohlhammer, 1886.

    Beschreibung des Oberamts Geislingen. Stuttgart and Tübingen: Cotta'schen, 1842.

    Beschreibung des Oberamts Gerabronn. Stuttgart and Tübingen: Cotta'schen, 1847.

    Beschreibung des Oberamts Heilbronn. Stuttgart: Kohlhammer, 1903.

    Beschreibung des Oberamts Leonberg. Stuttgart: Kohlhammer, 1930.

    Beschreibung des Oberamts Marbach. Stuttgart: Lindemann, 1866.

    Beschreibung des Oberamts Neckarsulm. Stuttgart: Kohlhammer, 1881.

    Beschreibung des Oberamts Neresheim. Stuttgart: Lindemann, 1872.

    Beschreibung des Oberamts Neuenbürg. Stuttgart: Karl Aue, 1860.

    Beschreibung des Oberamts Reutlingen. Stuttgart: Kohlhammer, 1893.

    Beschreibung des Oberamts Riedlingen. Stuttgart: Kohlhammer, 1923.

    Beschreibung des Oberamts Rottenburg. Stuttgart: Kohlhammer, 1900.

    Beschreibung des Oberamts Ulm. Stuttgart: Kohlhammer, 1925.

Ogiermann, Wilhelm. "Tauberbischofsheim im Mittelalter Urkundenforschung zu Kultur und Geschichte im Zeitraum von 800-1600." Tauberbischofsheim. Aus der Geschichte einer alten Amtsstadt. Tauberbischofsheim: Fränkische Nachrichten, 1955. Pp. 289-97.

Ott, Manfred. Lindau. Historischer Atlas von Bayern. Teil: Schwaben. München: Bayerische Historische Kommission, 1968.

Otto, Albrecht. Die evangelische Gemeinde Miltenberg und ihr erster Prediger. Schriften für das deutsche Volk, Vol. XXVIII. Halle: Verein für Reformationsgeschichte, 1891.

Peachey, Paul. Die Soziale Herkunft der Schweizer Täufer in der Reformationszeit. Schriftenreihe des Mennonitischen Geschichtsvereins, Vol. IV. Karlsruhe: Schneider, 1954.

Pfister, Johannes E. Denkwürdigkeiten der Württembergischen und Schwäbischen Reformationsgeschichte. Heilbronn: Renner, 1938.

Pfleger, Luzian. "Michael Hilsbach, ein oberrheinischer Schulmann des 16. Jahrhunderts." Zeitschrift für die Geschichte des Oberrheins, N.F. XX (1905), 252-59.

Plath, Helmut. "Verbreitungsgesetze in Brauch- und Wortgeographie Niedersachsens und angrenzender Gebiete." Neues Archiv für Niedersachsen, XV (1950), 51-67.

"Prediger-Historie der Reichsstadt Lindau im sechzehnten Jahrhundert." Historisch-Politische Blätter, LXII (1868), 497-530.

Pressel, Theodor. Anecdota Brentiana. Tübingen: Heckenhauer, 1868.

Rauch, Moriz von. "Johann Riesser, Heilbronner Reformationsbürgermeister." Historischer Verein Heilbronn, XVI (1925-1928), 11-23.

_____, ed. Urkundenbuch der Stadt Heilbronn. Stuttgart: Kohlhammer, 1916.

Rauscher, Julius. "Die ältesten Prädikaturen Württembergs." Blätter für Württembergische Kirchengeschichte, N.F. XXV (1921), 107-11.

_____. "Die Prädikaturen in Württemberg vor der Reformation." Württembergische Jahrbücher für Statistik und Landeskunde, II (1908), 152-211.

Redlich, Fritz. "Ideas. Their Migration in Space and Transmittal over Time." Kyklos, Internationale Zeitschrift für Sozialwissenschaft, VI (1953-54), 301-32.

"Die Reformation der Reichsstadt Biberach." Historisch-Politische Blätter, LVIII (1866), 720-37.

Reiff, Hans Jörg. Reformation und Verfassung in Reutlingen. Zulassungsarbeit. Tübingen: Institut für Landesgeschichte, 1970.

Reimer, Louis. "Reichsregierung und Reformation in Esslingen." Esslinger Studien, XI (1965), 226-40.

Reinwald, A. "Das Barfüsserkloster und die Stadtbibliothek in Lindau." Schriften des Vereins für die Geschichte des Bodensees, XVI (1887), 141-71.

Remling, Franz. Das Reformationswerk in der Pfalz. Mannheim: Götz, 1846.

_____. Geschichte der Bischöfe zu Speyer. Mainz: Kirchheim, 1881.

Richter, Friedrich. Zwei Schilderungen aus der Geschichte der ehemaligen Reichsstadt Bopfingen. Nördlingen: Beck'sche, 1862.

Roemer, Hermann. Geschichte der Stadt Bietigheim an der Enz. Stuttgart: Kohlhammer, 1956.

Rogers, Everett M. Diffusion of Innovations. New York: The Free Press, 1962.

Rohde, Hans Wilh. Evangelische Bewegung und Katholische Restauration im österreichischen Breisgau. Dissertation, Freiburg, 1957.

Rohling, Eugen. Die Reichsstadt Memmingen. Dissertation, München, 1864.

Rosenkranz, Albert. Der Bundschuh. Heidelberg: Winters, 1927.

Rösler, Hans. Geschichte und Strukturen der evangelischen Bewegung im Bistum Freising 1520-1571. Einzelarbeiten aus der Kirchengeschichte Bayerns, Vol. XLII. Nürnberg: Verein für Bayerische Kirchengeschichte, 1966.

Roth, F. "Otto Brunfels." Zeitschrift für die Geschichte des Oberrheins, IX (1894), 284-320.

Rublack, Hans Christoph. Die Einführung der Reformation in Konstanz von den Anfängen bis zum Abschluss 1531. Veröffentlichung des Vereins für Kirchengeschichte in der evangelischen Landeskirche Badens, Vol. XXVI. Lahr: Schauenburg, 1971.

Rückert, Hanns. "Die Bedeutung der württembergischen Reformation für den Gang der deutschen Reformationsgeschichte." Blätter für Württembergische Kirchengeschichte, N.F. XXXVIII (1934), 267-80.

Schaab, Meinrad. "Strassen und Geleitswege zwischen Rhein, Neckar und Schwarzwald im Mittelalter und der früheren Neuzeit." Jahrbuch für Statistik und Landeskunde von Baden-Württemberg, IV (1958), 54-75.

Schattenmann, Paul. Die Einführung der Reformation in die ehemalige Reichsstadt Rothenburg ob der Tauber, 1520-1580. Einzelarbeiten aus der Kirchengeschichte Bayerns, Vol. VII. Nürnberg: Verein für Bayerische Kirchengeschichte, 1928.

Schenck, Emil. Simprecht Schenck. Darmstadt: Selbstverlag, 1938.

Schiess, Traugott, ed. Briefwechsel der Brüder Ambrosius und Thomas Blaurer. Freiburg/Br.: Fehsenfeld, 1908-1912.

Schlenck, Wolfgang. Die Reichsstadt Memmingen und die Reformation. Memminger Geschichtsblätter, Jahresheft 1968. Memmingen: Heimatpflege, 1969.

Schmid, Wolfgang. Bönnigheim. Stadt und Ganerbiat. Zulassungsarbeit. Tübingen: Institut für Landesgeschichte, 1969.

Schmidt, Kurt Dietrich. "Die konfessionelle Gestaltung Deutschlands." Theologische Literaturzeitung, LXXVII (1952), 129-42.

Schneider, J. "Ein Brief M. Bucers an den Ritter Hans Landschad." Beiträge zur Hessischen Kirchengeschichte, III (1908), 103-16.

Schnellbögl, Fritz. "Die fränkischen Reichsstädte." Zeitschrift für Bayerische Landesgeschichte, XXXI (1968), 455-66.

Schöffler, Herbert. Die Reformation. Bochum: Langendreer, 1936.

Schöller, Peter. "Kräfte und Konstanten historisch-geographischer Raumbildung." Festschrift für Franz Petri zum 65. Geburtstag. Bonn: Röhrscheid, 1970. Pp. 476-84.

———. "Stadt und Einzugsgebiet. Ein geographisches Forschungsproblem und seine Bedeutung für Landeskunde, Geschichte und Kulturraumforschung." Studium Generale, X (1957), 602-12.

Schön, Theodor. "Beiträge zur Reformationsgeschichte Württembergs." Blätter für Württembergische Kirchengeschichte, VIII (1893), 77-78, 95-96.

———. "Meister Martin von Uhingen als Prediger in Rottenburg a. N." Reutlinger Geschichtsblätter, XIII (1902), 30.

———. "Wappenträger in Reutlingen." Reutlinger Geschichtsblätter, XVII (1906), 192-96.

Schornbaum, Karl. "Zur Reformation im Markgrafentum Brandenburg." Beiträge zur Bayerischen Kirchengeschichte, IX (1903), 26-35.

———. "Zur religiösen Stellung der Stadt Ansbach in den ersten Jahren der Reformation." Beiträge zur Bayerischen Kirchengeschichte, VII (1901), 146-66.

Schröder, Alfred. "Der Humanist Veit Bild, Mönch bei St. Ulrich." Zeitschrift des Historischen Vereins für Schwaben und Neuburg, XX (1892), 173-227.

Schröder, Richard, and Köhne, Karl, eds. Oberrheinische Stadtrechte. Heidelberg: Winter, 1898.

Schultze, Alfred. "Stadtgemeinde und Kirche im Mittelalter." Festschrift für Rudolf Sohm. Leipzig: Duncker & Humblot, 1914. Pp. 103-42.

———. Stadtgemeinde und Reformation. Recht und Staat in Geschichte und Gegenwart, Vol. XI. Tübingen: Mohr, 1918.

Schulze, Albert. Bekenntnisbildung und Politik Lindaus im Zeitalter der Reformation. Kirchengeschichte Bayerns, Vol. III. München: Kaiser, 1971.

Schuster, Otto. Kirchengeschichte von Stadt und Bezirk Esslingen. Stuttgart: Calwer, 1946.

Schwiebert, Ernest. "The Reformation from a New Perspective." Church History, XVII (1948), 3-31.

Sievers, Wilhelm. Über die Abhängigkeit der jetzigen Confessionsverteilung in Südwestdeutschland von den früheren Territorialgrenzen. Göttingen: Peppmüller, 1884.

Sigel, Christian. Das evangelische Württemberg. Seine Kirchenstellen und Geistlichen von der Reformation bis auf die Gegenwart. Gebersheim: Selbstverlag, 1938. Typewritten example at the Württembergische Landesbibliothek, Stuttgart.

Simon, Matthias. Ansbachisches Pfarrerbuch. Einzelarbeiten aus der Kirchengeschichte Bayerns, Vol. XXVIII. Nürnberg: Verein für Bayerische Kirchengeschichte, 1957.

_____. Evangelische Kirchengeschichte Bayerns. München: Müller, 1942.

_____. "Johann Rurer." Religion in Geschichte und Gegenwart, Vol. V. Tübingen: Mohr, 1956-1965. Pp. 1222-23.

_____. Pfarrerbuch der Reichsstädte Dinkelsbühl, Schweinfurth, Weissenburg in Bayern und Windsheim. Einzelarbeiten aus der Kirchengeschichte Bayerns, Vol. XXXIII. Nürnberg: Verein für Bayerische Kirchengeschichte, 1962.

_____. "Zur Reformationsgeschichte der Grafschaft Wertheim." Zeitschrift für Bayerische Kirchengeschichte, XXIX (1960), 121-44.

Social Science Research Council. The Social Sciences in Historical Study. Bulletin 64. New York, 1954.

Speh, Johann. Beiträge zur Reformation des oberen Neckargebietes. Dissertation, Tübingen, 1920.

Staehelin, Ernst. "Oecolompadiana." Basler Zeitschrift für Geschichte und Altertumskunde, LXV (1965), 165-94.

Steichele, Anton. Das Bistum Augsburg. Augsburg: Schmid, 1864-1934.

Steitz, Heinrich. Geschichte der evangelischen Kirche in Hessen und Nassau. Marburg: Trautvetter & Fischer, 1961.

Stolz, Aloys. Geschichte der Stadt Pforzheim. Pforzheim: Stadtverlag, 1901.

Störmann, Anton. Die städtischen Gravamina gegen den Klerus am Ausgange des Mittelalters und in der Reformationszeit. Reformationsgeschichtliche Studien und Texte, Vols. XXIV-XXVI. Münster: Aschendorff, 1916.

Störmer, Wilhelm. "Obrigkeit und evangelische Bewegung." Würzburger Diözesan-Geschichtsblätter, XXXIV (1972), 109-27.

Stupperich, Robert, ed. Martin Bucers Deutsche Schriften. Gütersloh: Mohn, 1960.

Tiryakian, Edward A. "A Model of Societal Change and Its Lead Indicator." The Study of Total Societies. Ed. Samuel Z. Klausner. Garden City: Doubleday, 1967. Pp. 69-97.

Toepke, Gustav, ed. Die Matrikel der Universität Heidelberg von 1386-1662. Heidelberg: Winter, 1884-1893.

Uhlhorn, Gerhard. Urbanus Rhegius. Elberfeld: Friderichs, 1862.

Urner-Astholz, Hildegard, et al. Geschichte der Stadt Stein am Rhein. Bern: Haupt, 1957.

Veit, Andreas L. Kirche und Kirchenreform in der Erzdiözese Mainz im Mittelalter, der Glaubensspaltung und der beginnenden tridentischen Reformation, 1517 bis 1618. Erläuterungen und Ergänzungen zu Janssens Geschichte des deutschen Volkes, Vol. X, No. 3. Freiburg: Herder, 1920.

Vierordt, Karl F. Geschichte der Reformation im Grossherzogtum Baden. Karlsruhe: Braun, 1847.

Virck, Hans, ed. Politische Korrespondenz der Stadt Strassburg im Zeitalter der Reformation. Strassburg: Trübner, 1882.

Vochezer, Josef. Geschichte von Waldburg. Kempten: Kösel'schen, 1900.

Wackernagel, Hans Georg, ed. Die Matrikel der Universität Basel. Basel: Universitätsbibliothek, 1951-56.

Wackernagel, Rudolf. Geschichte der Stadt Basel. Basel: Helbing & Lichtenhahn, 1968.

Wagner, Emil. "Die Reichsstadt Schwäbisch Gmünd in den Jahren 1523-1525." Württembergische Vierteljahreshefte für Landesgeschichte, N.F. II (1879), 26-33.

Walther, Heinrich. "Bernhard Besserer und die Politik der Reichsstadt Ulm während der Reformation." Ulm und Oberschwaben, XXVII (1930), 1-69.

Weber, Max. General Economic History. New York: Collier, 1961.

_____. The Theory of Social and Economic Organization. Ed. Talcott Parsons. Glencoe: Free Press, 1964.

Weidner, Karl. Die Anfänge der staatlichen Wirtschaftspolitik in Württemberg. Darstellungen aus der württembergischen Geschichte, Vol. XXI. Stuttgart: Kohlhammer, 1931.

Weigel, Hellmut. Die Deutschordenskomturei in Rothenburg o.T. im Mittelalter. Quellen und Forschungen zur Bayerischen Kirchengeschichte, Vol. VI. Leipzig: Deichert, 1921.

Weigel, M. "Der erste Reformationsversuch in der Reichsstadt Kaufbeuren und seine Niederwerfung." Beiträge zur Bayerischen Kirchengeschichte, XXI (1914), 145-56.

Weyermann, Albrecht. "Die Bürger in Ulm der zwinglischen Confession zugetan." Tübinger Zeitschrift für Theologie, XII (1830), 142-53.

Wipf, Jacob. Reformationsgeschichte der Stadt und Landschaft Schaffhausen. Zürich: Füssli, 1929.

_____. "Zwingli's Beziehungen zu Schaffhausen." Zwingliana, V (1929-1933), 11-41.

Weller, Karl. Württembergische Vergangenheit. Stuttgart: Kohlhammer, 1932.

Wulz, Gustav. "Caspar Kantz." Lebensbilder aus Bayerisch-Schwaben, Vol. IV. München: Huber, 1955. Pp. 100-119.

Wunder, Gerd. "Georg Widmann (1486-1560) und Johann Herolt (1490-1562), Pfarrer und Chronisten." Lebensbilder aus Schwaben und Franken, VII (1960), 41-51.

"Württembergisches aus römischen Nuntiaturberichten 1521-32." Blätter für Württembergische Kirchengeschichte, VIII (1893), 78-81.

Zeeden, Ernst W. "Grundlagen und Wege der Konfessionsbildung in Deutschland im Zeitalter der Glaubenskämpfe." Historische Zeitschrift, CLXXXV (1958), 249-99.

Zell, Joseph. "Andreas Althammer als Altertumsforscher." Württembergische Vierteljahreshefte für Landesgeschichte, N.F. XIX (1910), 429-44.

Zeller, Josef. "Aus dem ersten Jahrhundert der gefürsteten Propstei Ellwangen (1460-1560)." Württembergische Vierteljahreshefte für Landesgeschichte, N.F. XVII (1908), 278-82.

Zoepfl, Friedrich. Das Bistum Augsburg und seine Bischöfe im Reformationsjahrhundert. Augsburg: Winfried, 1969.

"Zur Geschichte des Bergbaues von Durlach von 1439 bis 1532." Zeitschrift für die Geschichte des Oberrheins, I (1850), 43-48.

Unpublished Material

Herdsteuerlisten 1525. Haupstaatsarchiv Stuttgart. Bestand A 54a, Lists 19, 20, 21, 22, 23, 24, 25, 26, 27, 28, 29, 30, 31, 32, 33, 34, 35, 36, 37, 38, 39, 40, 41, 42, 43, 44, 45, 46, 47, 48, 49, and 50.

THE UNIVERSITY OF CHICAGO
DEPARTMENT OF GEOGRAPHY
RESEARCH PAPERS (Lithographed, 6×9 Inches)

(Available from Department of Geography, The University of Chicago, 5828 S. University Ave., Chicago, Illinois 60637. Price: $6.00 each; by series subscription, $5.00 each.)

106. SAARINEN, THOMAS F. *Perception of the Drought Hazard on the Great Plains* 1966. 183 pp.
107. SOLZMAN, DAVID M. *Waterway Industrial Sites: A Chicago Case Study* 1967. 138 pp.
108. KASPERSON, ROGER E. *The Dodecanese: Diversity and Unity in Island Politics* 1967. 184 pp.
109. LOWENTHAL, DAVID, et al. *Environmental Perception and Behavior.* 1967. 88 pp.
110. REED, WALLACE E. *Areal Interaction in India: Commodity Flows of the Bengal-Bihar Industrial Area* 1967. 210 pp.
112. BOURNE, LARRY S. *Private Redevelopment of the Central City: Spatial Processes of Structural Change in the City of Toronto* 1967. 199 pp.
113. BRUSH, JOHN E., and GAUTHIER, HOWARD L., JR. *Service Centers and Consumer Trips: Studies on the Philadelphia Metropolitan Fringe* 1968. 182 pp.
114. CLARKSON, JAMES D. *The Cultural Ecology of a Chinese Village: Cameron Highlands, Malaysia* 1968. 174 pp.
115. BURTON, IAN; KATES, ROBERT W.; and SNEAD, RODMAN E. *The Human Ecology of Coastal Flood Hazard in Megalopolis* 1968. 196 pp.
117. WONG, SHUE TUCK. *Perception of Choice and Factors Affecting Industrial Water Supply Decisions in Northeastern Illinois* 1968. 96 pp.
118. JOHNSON, DOUGLAS L. *The Nature of Nomadism* 1969. 200 pp.
119. DIENES, LESLIE. *Locational Factors and Locational Developments in the Soviet Chemical Industry* 1969. 285 pp.
120. MIHELIC, DUSAN. *The Political Element in the Port Geography of Trieste* 1969. 104 pp.
121. BAUMANN, DUANE. *The Recreational Use of Domestic Water Supply Reservoirs: Perception and Choice* 1969. 125 pp.
122. LIND, AULIS O. *Coastal Landforms of Cat Island, Bahamas: A Study of Holocene Accretionary Topography and Sea-Level Change* 1969. 156 pp.
123. WHITNEY, JOSEPH. *China: Area, Administration and Nation Building* 1970. 198 pp.
124. EARICKSON, ROBERT. *The Spatial Behavior of Hospital Patients: A Behavioral Approach to Spatial Interaction in Metropolitan Chicago* 1970. 198 pp.
125. DAY, JOHN C. *Managing the Lower Rio Grande: An Experience in International River Development* 1970. 277 pp.
126. MAC IVER, IAN. *Urban Water Supply Alternatives: Perception and Choice in the Grand Basin, Ontario* 1970. 178 pp.
127. GOHEEN, PETER G. *Victorian Toronto, 1850 to 1900: Pattern and Process of Growth* 1970. 278 pp.
128. GOOD, CHARLES M. *Rural Markets and Trade in East Africa* 1970. 252 pp.
129. MEYER, DAVID R. *Spatial Variation of Black Urban Households* 1970. 127 pp.
130. GLADFELTER, BRUCE. *Meseta and Campiña Landforms in Central Spain: A Geomorphology of the Alto Henares Basin* 1971. 204 pp.
131. NEILS, ELAINE M. *Reservation to City: Indian Urbanization and Federal Relocation* 1971. 200 pp.
132. MOLINE, NORMAN T. *Mobility and the Small Town, 1900–1930* 1971. 169 pp.
133. SCHWIND, PAUL J. *Migration and Regional Development in the United States, 1950–1960* 1971. 170 pp.
134. PYLE, GERALD F. *Heart Disease, Cancer and Stroke in Chicago: A Geographical Analysis with Facilities Plans for 1980* 1971. 292 pp.
135. JOHNSON, JAMES F. *Renovated Waste Water: An Alternative Source of Municipal Water Supply in the U.S.* 1971. 155 pp.
136. BUTZER, KARL W. *Recent History of an Ethiopian Delta: The Omo River and the Level of Lake Rudolf* 1971. 184 pp.
137. HARRIS, CHAUNCY D. *Annotated World List of Selected Current Geographical Serials in English, French, and German* 3rd edition 1971. 77 pp.
138. HARRIS, CHAUNCY D., and FELLMANN, JEROME D. *International List of Geographical Serials* 2nd edition 1971. 267 pp.
139. MC MANIS, DOUGLAS R. *European Impressions of the New England Coast, 1497–1620* 1972. 147 pp.
140. COHEN, YEHOSHUA S. *Diffusion of an Innovation in an Urban System: The Spread of Planned Regional Shopping Centers in the United States, 1949–1968* 1972. 136 pp.

141. MITCHELL, NORA. *The Indian Hill-Station: Kodaikanal* 1972. 199 pp.
142. PLATT, RUTHERFORD H. *The Open Space Decision Process: Spatial Allocation of Costs and Benefits* 1972. 189 pp.
143. GOLANT, STEPHEN M. *The Residential Location and Spatial Behavior of the Elderly: A Canadian Example* 1972. 226 pp.
144. PANNELL, CLIFTON W. *T'ai-chung, T'ai-wan: Structure and Function* 1973. 200 pp.
145. LANKFORD, PHILIP M. *Regional Incomes in the United States, 1929–1967: Level, Distribution, Stability, and Growth* 1972. 137 pp.
146. FREEMAN, DONALD B. *International Trade, Migration, and Capital Flows: A Quantitative Analysis of Spatial Economic Interaction* 1973. 202 pp.
147. MYERS, SARAH K. *Language Shift Among Migrants to Lima, Peru* 1973. 204 pp.
148. JOHNSON, DOUGLAS L. *Jabal al-Akhdar, Cyrenaica: An Historical Geography of Settlement and Livelihood* 1973. 240 pp.
149. YEUNG, YUE-MAN. *National Development Policy and Urban Transformation in Singapore: A Study of Public Housing and the Marketing System* 1973. 204 pp.
150. HALL, FRED L. *Location Criteria for High Schools: Student Transportation and Racial Integration* 1973. 156 pp.
151. ROSENBERG, TERRY J. *Residence, Employment, and Mobility of Puerto Ricans in New York City* 1974. 230 pp.
152. MIKESELL, MARVIN W., editor. *Geographers Abroad: Essays on the Problems and Prospects of Research in Foreign Areas* 1973. 296 pp.
153. OSBORN, JAMES. *Area, Development Policy, and the Middle City in Malaysia* 1974. 273 pp.
154. WACHT, WALTER F. *The Domestic Air Transportation Network of the United States* 1974. 98 pp.
155. BERRY, BRIAN J. L., et al. *Land Use, Urban Form and Environmental Quality* 1974. 464 pp.
156. MITCHELL, JAMES K. *Community Response to Coastal Erosion: Individual and Collective Adjustments to Hazard on the Atlantic Shore* 1974. 209 pp.
157. COOK, GILLIAN P. *Spatial Dynamics of Business Growth in the Witwatersrand* 1975. 143 pp.
158. STARR, JOHN T., JR. *The Evolution of Unit Train Operations in the United States: 1960–1969—A Decade of Experience* 1975.
159. PYLE, GERALD F. *The Spatial Dynamics of Crime* 1974. 220 pp.
160. MEYER, JUDITH W. *Diffusion of an American Montessori Education* 1975. 109 pp.
161. SCHMID, JAMES A. *Urban Vegetation: A Review and Chicago Case Study* 1975.
162. LAMB, RICHARD. *Metropolitan Impacts on Rural America* 1975.
163. FEDOR, THOMAS. *Patterns of Urban Growth in the Russian Empire during the Nineteenth Century* 1975.
164. HARRIS, CHAUNCY D. *Guide to Geographical Bibliographies and Reference Works in Russian or on the Soviet Union* 1975. 496 pp.
165. JONES, DONALD W. *Migration and Urban Unemployment in Dualistic Economic Development* 1975.
166. BEDNARZ, ROBERT S. *The Effect of Air Pollution on Property Value* 1975. 118 pp.
167. HANNEMANN, MANFRED. *The Diffusion of the Reformation in Southwestern Germany, 1518-1534* 1975.
168. SUBLETT, MICHAEL D. *Farmers on the Road. Interfarm Migration and the Farming of Noncontiguous Lands in Three Midwestern Townships, 1939-1969* 1975. 228 pp.
169. STETZER, DONALD FOSTER. *Special Districts in Cook County: Toward a Geography of Local Government* 1975. 189 pp.
170. EARLE, CARVILLE V. *The Evolution of a Tidewater Settlement System: All Hallow's Parish, Maryland, 1650–1783* 1975. 249 pp.
171. SPODEK, HOWARD. *Urban-Rural Integration in Regional Development: A Case Study of Saurashtra, India—1800–1960* 1975.
172. COHEN, YEHOSHUA S. and BERRY, BRIAN J. L. *Spatial Components of Manufacturing Change* 1975.
173. HAYES, CHARLES R. *The Dispersed City: The Case of Piedmont, North Carolina* 1975.
174. CARGO, DOUGLAS B. *Solid Wastes: Factors Influencing Generation Rates* 1975.
175. GILLARD, QUENTIN. *Incomes and Accessibility. Metropolitan Labor Force Participation, Commuting, and Income Differentials in the United States, 1960–1970* 1975.
176. MORGAN, DAVID J. *Patterns of Population Distribution: A Residential Preference Model and Its Dynamic* 1975.
177. STOKES, HOUSTON H.; JONES, DONALD W. and NEUBURGER, HUGH M. *Unemployment and Adjustment in the Labor Market: A Comparison between the Regional and National Responses* 1975. 135 pp.